THE STARRY ROOM

NAKED EYE ASTRONOMY
IN THE INTIMATE UNIVERSE

FRED SCHAAF

FOREWORD BY CHET RAYMO

DOVER PUBLICATIONS, INC.
MINEOLA, NEW YORK

Dedication

MY FATHER died during the time of this book's writing. *The Starry Room* is dedicated to him and to my mother.

Credits

Illustrations: Doug Myers
Technical Drawings: Guy Ottewell

Bibliographical Note

This Dover edition, first published in 2002, is an unabridged and slightly corrected and updated republication of the work originally published in 1990 by John Wiley & Sons, Inc., New York. A new Introduction to the Dover Edition has been specially prepared for this reprint by the author.
PLEASE NOTE: Like the original, not every illustration is provided with a caption, as was the author's intention.

Library of Congress Cataloging-in-Publication Data

Schaaf, Fred.
   The starry room : naked eye astronomy in the intimate universe / Fred Schaaf ; foreword by Chet Raymo.
      p. cm.
   Originally published: New York : Wiley, c1988.
   Includes bibliographical references and index.
   ISBN 0-486-42553-3 (pbk.)
    1. Astronomy—observers' manuals. I. Title.

QB64 .S43 2002
522—dc21

                                       2002073720

Manufactured in the United States of America
Dover Publications, Inc., 31 East 2nd Street, Mineola, N.Y. 11501

FOREWORD

There are people with gifted palates who become winetasters. There are people with perfect pitch who become piano tuners. There are people with sensitive noses who work for perfumeries. Fred Schaaf has the gift of sight; he writes about the night sky.

The naturalist John Burroughs said that two things are necessary for the art of seeing: knowledge and love. Schaaf has both in abundance. He is an amateur astronomer with the breadth of knowledge of the professional. And he knows some things the professionals do not, like optimum ways of seeing with the naked eye. He has seen things the professionals say are too faint to see without optical aid—7th magnitude stars and a partial penumbral eclipse of the moon are examples—and he teaches us how to do it. Schaaf's knowledge is acquired from a long and intimate acquaintance with the night; some of the things you will learn in this book you will find no place else.

If Schaaf is an amateur astronomer, he is an amateur in the original meaning of the word—one who loves. I read somewhere that the root of the Latin word for love—*am*—derives from baby-talk. Like yum-yum or mmmm! it was an expression of delight. *The Starry Room* is the record of one man's quest for the *am*, those special moments when we experience a thing not experienced (by us) before—an eyelash-thin moon, a fire-green shooting star, a comet, an eclipse—and suddenly we know the world in a new and more perfect way.

Fred Schaaf's love and knowledge of the night are infectious. I have been looking at the sky and reading the astronomical literature for more than thirty years; what I found in this book is new and fresh. Energy and curiosity shine like starlight on every page. Schaaf's vivid descriptions of celestial events made me wish I had been with him the night of the especially dark lunar eclipse of December 30, 1982, as the "exquisite chiselry" of the lunar surface was "overwhelmed utterly by a wave of inky darkness." Or the night at Dyers Cove when meteors rained down from the heavens and a thunderstorm on the southern horizon "flickered a mystical accompaniment." Or the time he saw the fireball "like a mass of green fire" trailing tongues of flame. The book is full of such moments, passionately experienced, beautifully described. It is also full of the kind of information that makes it possible for you and me to see such wonders and to understand what is seen.

Perhaps it is wrong to call Schaaf an amateur astronomer. The adjective has lost something of its original stature. Better to call him a naturalist of the night. This book places him squarely in the naturalist tradition, the tradition of Thoreau and Muir and Burroughs, a tradition that bridges the gap between the sciences and the humanities. In the introduction to *Desert Solitaire,* the naturalist Edward Abbey says: "When traces of blood begin to mark your trail you'll see something, maybe." It is the traces of blood that distinguish the naturalist from the professional scientist. It is the traces of blood that make this book an irresistible trail to follow.

What makes this book important is the *intensity* of Schaaf's experience. After I had finished reading about half of it I felt compelled to leave my chair and go out and look at the night. Schaaf makes you feel that if you just look up at the sky you will see something spectacular, and—you know—he is right.

—Chet Raymo

ACKNOWLEDGMENTS

My first thanks go to Doug Myers for his excellent work in providing the majority of this book's illustrations. I am especially delighted by Doug's scratchboard art, which besides being marvelously done strikes me as being very much in keeping with the spirit of my text. I also am grateful to the other provider of illustrations, Guy Ottewell. In addition to permitting my use of diagrams from a few of his past works, Guy at my request computer-plotted several new ones with skill and artistry especially for this book.

David Meisel was a tremendous help on meteors and gave generously of his time to help me solve the mysteries of the great 1982 New Jersey fireball. I could not have provided him with so many useful observations without the supportive readers of my "South Jersey Skies" column in the *Atlantic City Press.*

My friend Steve Albers supplied important background information; I also wish to acknowledge and thank him for inventing the kind of "indoor sky" described in Chapter 9.

I am indebted also to the following people for knowledge and materials in specific areas: Dave Crawford and Rick Kurczewski (light pollution); Ruth Freitag, Joe Laufer, and Charles Morris (Halley's Comet); and Robert C. Victor (planets).

My final editor, David Sobel, and managing editor, Corinne McCormick, at Wiley, brought this book to completion skillfully, punctually—and warmly. Barbara Conover and Publishing Synthesis performed production aspects swiftly and agilely. Last but not least, my thanks go to Mary Kennan, the editor originally interested in this book who guided the manuscript all the way from inception right up until its final stages.

The quotations on p. 107 and p. 135 are from *The Poems of William Butler Yeats—A New Edition,* edited by Richard Finneran (New York: Macmillan, 1983). The quotations on pp. 125–26 are from an interview in *Meteor News* conducted by John West and David Swann, and are reproduced here with their kind permission. The quotations on pp. 215–17 are from "On Fairy Stories" in *The Tolkien Reader* (copyright © 1966 by J.R.R.

Tolkien), and are reprinted by permission of Houghton Mifflin Company.

Last and most of all, I wish to thank my wife Mamie. Without her support, encouragement, and efforts, *The Starry Room* would not have been accomplished.

INTRODUCTION TO THE DOVER EDITION

Of all the books I've written, *The Starry Room* is the one for which readers seem to have the greatest affection. Canadians Guy Boily and Lucie Granger wrote in their article in *Sky and Telescope* (pp. 551–553, November 1990) that reading this book gave them the boost they needed to build their own log-cabin observatory. Todd Fouts found a copy of *The Starry Room* in a small bookshop in Tokyo, where he lives under heavy light pollution. He says that he frequently rereads parts of the book, and is brought to tears as it helps him recall the night sky in Ohio where he grew up—a dark sky filled with stars and the sound of migrating geese. After reading *The Starry Room*, a mentally challenged man in Tennessee conceived an abiding love for astronomy, and he longs as much as I do to live to see the next supernova (see Chapter 11). Many other people have told me that they have found inspiration in this passionate book.

It is therefore with great pleasure that I introduce the Dover reprint of *The Starry Room*.* Since much of *The Starry Room* is a record of the wonder of my own actual astronomical observations, it doesn't really require a great deal of informational updating. Along with a very few other changes, my Dover editor, Joslyn Pine, and I have updated the bibliography and reworked it into a "Suggested Reading/Resource List." The publications *Sean News Bulletin* and *Meteor News* mentioned in Chapters 6 and 7 no longer exist in the same form or name, so you'll need to refer to this list on page 257 for the comparable early twenty-first century sources.

Of course, in the fifteen years since I wrote *The Starry Room*, I have had numerous further adventures with many of the astronomical objects and sky phenomena recounted in these chapters. But these new experiences would be better served in a sequel to *The Starry Room*. Will I write such a sequel? The response from you, good readers, may play a role in answering this question positively. If you would like to see a sequel become a reality, or indeed if you have any praise or criticism for the current book, I ask you to write to me at fschaaf@aol.com. Whether I hear from you or not, I hope this edition of *The Starry Room* will bring you pleasure and get you looking up at the sky.

FRED SCHAAF
June 2002

---

*If you like this book, there is another title of mine you will probably enjoy, also available from Dover Publications, Inc.—*Wonders of the Sky* (0-486-24402-4), my very first book.

CONTENTS

INTRODUCTION TO THE DOVER EDITION   vii

INTRODUCTION   1

1   The Secrets of Seeing   6

2   Year of the Mandala and Monarch Moon Eclipses   20

3   The Many Suns of the Daytime Sky   42

4   100 Rainbows   50

5   A Night at Meteor Cove   70

6   Flight   82

7   Go to Innisfree   106

8   All the Worlds in My Window   136

9   Making the Indoor Sky   156

10   The Powers of Vision   165

11   The Next Supernova   192

12   The End of the Stars (Not One Child in Ten)   204

13   The Best and Worst of Returns   226

14   The Fate and Meaning of Halley's Comet   236

15   Walking to the Stars   250

SUGGESTED READING/RESOURCE LIST   257

GLOSSARY   258

INDEX   260

INTRODUCTION

As THE GLARE from city lights continues to spread and worsen (for the most part unnecessarily and avoidably), the age-old splendor of the night sky is in danger of being lost and forgotten. We must not forget it, or let it be spoiled. But what is strange and unfortunate is that even many astronomers and astronomy books are contributing to the growing amnesia. Their concern is with only objects, only in the context of outer space (not the sky), and only as revealed by the automated means with which, admittedly, professionals can often uncover the most facts about these objects. Such an approach has helped bury away, at the time we need it most, the old delight and wonder of standing under and observing directly the beauties of the starry sky.

This book is a step back out under that sky, a return to its wonders. It certainly is not a rejection of the new instruments and techniques. The knowledge they have brought us now enables us to look at celestial objects with greater appreciation and wonder than ever, comprehending better than before their marvelous nature and still richer relations. But the fact remains that we need to care the best and most justly we can about the things and not just about our accumulation of statistics on them or our uses of them in theoretical schemes. Without our perceptions of the sky and resultant feelings of awe, delight, and connection, even our most strictly scientific conceptions of outer space themselves will become starved, and the science of astronomy suffer. But even that would be a small tragedy compared with the spiritual bankruptcy of ourselves which losing the sky and stars would mean.

We should not forget ourselves in our role as appreciators of these heavens we study. There is no feeling of insignificance or meaninglessness for anyone who is an active participant in this appreciation, which not only involves us with the cosmos but makes us intimate with it as only friends or lovers can be. And only through the medium of not just our senses, thought, and feelings but also through a *sky* can we have this involvement. We are very lucky that wherever we go—even to space—there is always a sky through which to make contact with that greater universe out there.

But perhaps we should not talk about the sky as medium and the universe alone as place. Perhaps they are better envisioned together and in relation with us, all at the same time, as a place—not a simply vast, abstracted void, but a place in which to be and see and relate. We cannot cram all the wonder of the starry sky and the delight it occasions into a mere universe of 30 or more billion light-years' (current) diameter and 10 or 15 billion years' (current) age. But perhaps we can fit it all into a room. . . .

This book is most essentially a collection of essays about adventures in the heart of astronomy—that heart which knows it is in the Starry Room. What do I mean by this metaphor of the title? A room is a place for keeping, a place of some considerable permanence, a place with a purpose (at the very least our own purpose)—even if that purpose is "only" living and all that which goes with living. A room is also a place in which some kind of intimacy is possible. A stretch of the imagination may be required to see how the vast, supposedly four-dimensional, maybe open and maybe closed chamber of our universe is in these ways like a room—especially how the great starry universe with its people-dwarfing gaseous rolling fires and still more enormous, cold, dark distances can be intimate to us. But it is. And it can be.

My purpose in this book is to describe many of the beautiful features of this physically greatest and most enduring of rooms and also to show how the best way to learn them is through your own personal, intimate discoveries of them. In the most ultimate sense, there is no true replacement for direct observation in astronomy. Every essay in this book bears witness to that truth—often through my own eyes, through my own personal (private yet universal!) experience of celestial wonders.

Most of these essays also demonstrate that you do not necessarily need telescopes and other equipment, or perfect sky conditions, or a long period of special training to start seeing marvelous sights in the heavens which most other people miss. Your own naked eyes may be all you need for the all-important beginning—provided you also have some desire and understanding, such as I hope with this book to spark and supply. (I believe that even readers of long experience in astronomy and skywatching will find here—especially in chapters on rainbows, halos, meteors, meteorites, supernovae, planetary conjunctions, light pollution, and sky painting—many exciting facts they have never encountered in *any* book before.) Telescopes can provide tremendous new levels and vistas for exploration as long as they are used with the knowledge and love of the heavens which usually comes best from first doing naked-eye observing. We should remember that genuine enthusiasm and understanding must be a part of that instrument behind the eyes which is more important than any you can put in front of them!

I hope that readers will be far more knowledgeable about a variety of astronomical subjects after finishing this book. But how much initial knowledge of astronomy do you need to understand the book? Practically none. Most technical terms are explained as they appear in the text, and the essays (chapters) can be read independently of one another (although there is much linking them, and together they certainly do produce a whole greater than the sum of its parts). A few terms or concepts which appear in many of the essays are defined in full in a small glossary at the back of the book. The beginner may wish to scan over that glossary before starting the adventure of the chapters.

I think it may not be just whimsical or useless to ask in what house this Starry Room is located. But if I may pass on that bold (some would say rash—or absurd) question, I can turn to a far less confident (but nevertheless productive) one: is the universe too big to be a "room"?

The bigness of the universe if invoked too often or too importantly becomes after all a small—that is to say, a petty—thing. We should remember that the Starry Room has many other qualities or properties than vastness, many equally or more arresting and wondrous.

Do we know the articles and parts of the Starry Room, let alone its entirety and full story, and know them so well that we can cease to be excited and discard or forget them? To think so is a very great folly. One of the fundamental principles which science accepts without rigorous (but nevertheless pretty good) proof is that its laws are truly universal, that our sampling of a bit of the universe here applies quite well to what of it is over there or beyond our direct reach. But even if this is true (to keep things thus actually requires a lot of ongoing redefining and restructuring), the other parts are not exactly the same. We seem to have no black holes or active supernovae or quasars, or many other things less violent but equally wondrous, in the area of space and time near us and our solar system. Even some of the closest parts of the Starry Room—planets, comets, meteorites crashing through our roofs—are perhaps more enigmatic than known.

If the universe is not just terribly big and old, and its history is not easily written off on the blackboard or in the textbook once and for all, then what is it? The Starry Room is vast and ancient indeed, but it is so much more. In this book you will encounter a number of its most exciting parts and events which roar and pierce and soar and chill with beauty—an infinitesimal sample of all it has to offer but a significant one nonetheless, I think, because the events are presented, as they were experienced, with an intense awareness of the Room and its atmosphere of wonder.

When you read about the storm held at bay and the peace-filled night of meteors like my hand filled with luminescent creatures; about the shadow-casting, fiercely burning, devastatingly silent fireball sailing over my head and fear-outstretched hand; about the glimpse I and I almost alone in the world had of Halley's greatest immensity of tail glimmering like a ghost 10,000 times the size of the Earth; about all the worlds of the solar system concentrated for the last time in centuries into one window—the kitchen window I had grown up looking out; about the moon striking me like lightning as it reappeared from darkest volcano-shadowed eclipse like a vast and mighty monarch butterfly by the old lighthouse which I hoped would still be standing to greet its next such incredibly rare migration . . . when you read about all these things, I hope a shiver or two will run through you as they did through me: a shiver of awareness, awesome and beautiful, racing out to the very reaches of that wondrous Room.

And I hope that this reading will help you rediscover the way to your own adventures, or more of your own adventures, in this place where the most distant starlight may become as intimate as our own thoughts and longings.

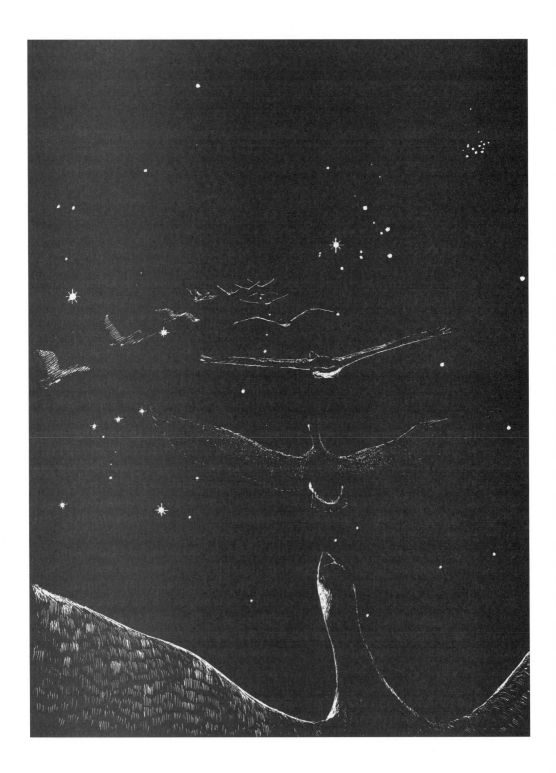

1

The Secrets of Seeing

IT WAS THE SOLE DARK HOUR between moonset and morning twilight on
October 26, 1985, and I was out at my parents' home in the country to
observe Halley's Comet. About ten seconds after I had first looked up, I
had seen one of the Orionid meteors, flaming streaks caused by tiny pieces
of dust which had been released by Halley's to glow in its tail dozens or
hundreds of its once-in-a-lifetime visits earlier. Orionid meteors are visible
every year around this time in October, but to see a few with their parent
comet in the sky near the part of Orion they fly out of was a chance seldom
available in all of history. Too bad that Halley was not yet visible to the
naked eye—though beautiful and intense-centered in my telescope.

I needed no other rare experience to make that quiet, dark hour memorable, but one of my reveries was broken suddenly when I heard a noise.
For a moment, then another, I did not recognize it, for I had never heard
it so loud and close in the still of the night. Then there was no doubt. Geese!
It was the sound of Canada geese. This group was weeks later than most
on the autumn migration. And they were *close*. North of me down the
wooded road a bit the neighbor's dog barked. The honking of the geese
was louder, sharper, resounding, filling everything. And yet (to my amazement) the rest of the world kept sleeping. Only I and the dog, it seemed,
were awake in all the world. I could not help but hope. On this night of
magical Halley sights, might I be granted a further vision?

All of my mind and body doubted. In twenty-five years of watching the
night sky I had looked for geese crossing the stars dozens of times when I
heard their lonely, heart-stirring call on high in the night. A few glimpses
of them near the moon, exciting but indistinct, always short of what I had
dreamed of, was the best that I ever got. That was all. So now my mind

and body doubted. But my body still scrambled, my head turned, I moved out from under trees to where something told me the right section of the constellations would be visible. And lifted to the starry heavens my eye. . . .

Nothing is so heroic as the eye. Not just the perfect or idealized eye in something like a Renaissance painting or Greek statue, but yours or mine, in something like its proper use. The truth of this is staring us from the face, yet how few people really know it! The catch of course, is that "proper use"—learning how to put your average yet potentially miraculous vision to it.

There are secrets to seeing. Some are simple, some not so simple. Some of them are physiological—and these I discuss much later in this book (in Chapter 11, "The Powers of Vision"). But most of the secrets are more a matter of knowledge and what we may call attitude or spirit.

One thing else is certain, too: there is no better place to learn the secrets of seeing than in the sky. Astronomy, of all sciences or nature studies, is the one which always has and always will depend most heavily on the eye. Even in our recent decades of handled moon rocks, "felt" Martian winds and quakes (and "tasted" Martian soil), "heard" radio waves from natural sources (really "seen" in a range beyond that of visible light), we recognize most of the universe as too vast and too far to ever be reached or studied with any of the senses but vision. This is true even assuming that we someday build faster-than-light spaceships and visit the stars—the universe is too big (and, in each place, too small) to be touched everywhere firsthand, though maybe not writ too large and small to be read with limited yet still wonderful comprehension. Even those great augmenters of the eye's powers, the telescope and photographic film, now joined impressively by electronic light detectors, even those are ultimately just the accoutrements of human vision. They do what they do for the eye.

But astronomy's already veteran, foreseeably continuing and naturally intimate relationship with the eye is not the best reason for you or me to learn the secrets of seeing in the heavens. The best reason is: if the eye is a hero—or can become one when we give it the right kind of chance— then the heavens are the eye's widest and often grandest field in which to adventure. Every night will bring to sight another wonder. And not even 25 years—or 250—will exhaust the new ones . . . like a line of geese among the stars.

Our initial set of secrets are those involving knowledge about the wheres and whens for looking.

There are times and situations to avoid if you want to observe specific kinds of sky sights. Amateur astronomers soon learn that a night of bright moon is not good for most kinds of astronomical observing because the night's other celestial objects are so much fainter, so overwhelmed by the flood of lunar radiance. There are a few possible exceptions. If you are learning the brightest stars and constellations, some experts suggest that the patterns are easier to pick out in a moon-washed or city (but not very big city) sky than when they are accompanied by a confusing welter of several thousand more faint stars in a dark, moonless country sky. Personally, I cannot help but feel that the struggle to find these stars and patterns amid so rich a profusion of fainter stars is more than made up for by the glorious experience of that profusion (not all bewilderment is bad!) and by the greater natural splendor of the bright stars and constellations seen as they should be—in a dark heavens. Be all that as it may, telescopic observations of bright planets do not necessarily suffer from moonlight, manlight, or bright twilight. As a matter of fact, the surface brightness (actually cloud brightness) of a planet like Venus is best seen when it does not contrast too strongly—a dazzle in the dark—with the background sky. Likewise, you may see more detail on Jupiter or Mars when the sky around them is not very dark.

Other than these and a few other exceptions, the moon—lovely though it be—is something to avoid if you want to observe other celestial objects well. Indeed, though the fact comes as a big surprise to the novice, a brightly moonlit night is not in many respects the best time for observing even the moon itself!

What is the problem with observing the moon with binoculars or telescope anytime around full moon? The problem is not the moon's great total brightness then; it is the angle at which sunlight is striking lunar features—most perpendicularly to the surface and thus with hardly any shadows to outline craters, mountain ranges, valleys, and other topography. A few features can be seen properly *only* around full moon—for instance, the spectacular "rays" which radiate in surface streaks of ejected dust hundreds of miles long (and longer) from some of the moon's younger, fresher craters. The rays have no real depth or height and so cast no shadows. For best views of all those lunar features which do cast shadows, the time to look is when they are near the moon's "terminator." The terminator is the line separating light and dark, day from night, on the moon—the line along which astronauts on the moon would be witnessing sunrise or sunset. It is where shadows are longest. Near the terminator you can often catch sight of a lunar peak's summit shining as a bright speck surrounded by darkness (a tiny detached piece of moonlight!) or of a crater looking like a ring of light exquisitely chiseled around a hole of profound darkness. Devoted lunar observers want to see this fair satellite's face in all its lights and guises, of course, but beginners can get some of the easiest and most spectacular views by looking at the right time and place. The moon around first quarter (conveniently high in the early-evening sky) is especially good to look at, with not yet so much straight-on sunlight and the terminator lying across particularly rough highlands and rugged numerous craters. With a telescope, the sunrise's advance over this dramatic lunar landscape can be appreciated not just from night to night but even from hour to hour.

Even though having skies free from bright moon or man-made "light pollution" is not quite always important, you might think that the clearness of the sky is. But for certain kinds of astronomical observing, the darkest and clearest nights can nevertheless be poor! These are the kinds of observing in which extreme steadiness and sharpness of image are crucial. After a cold front has roared through your area the atmosphere may be at its most transparent, but you may not see details on planets or split the close-together members of a double star system with a telescope. Why not? Because the atmosphere you are looking through in such a weather situation is likely to be turbulent. The effect of the atmosphere's turbulence on the sharpness and steadiness of images is what astronomers call the "seeing"— good "seeing" if the atmosphere is steady and calm, bad "seeing" if it is especially turbulent. The turbulence need not be windiness anywhere near ground level—bad "seeing" is more often and largely a result of irregular air flow at higher altitudes in the atmosphere above you. A rough naked-eye guide to the "seeing" is provided by how strongly stars twinkle: the more twinkling (though it is pretty in itself), the worse the "seeing." Since starlight must pass through a longer pathway of air down low in the sky, stars always twinkle a lot down near the horizon. Strong twinkling of a star high in the sky is an indication of very bad "seeing." The night may be startlingly clear, perfect for glimpsing faint meteors and splendor of constellations with the naked eye or delicate, dim galaxies and faint star clusters with the telescope (though not precise detail in them)—but you will probably not split close double stars or see much more than wavering glimpses of bands on otherwise featureless Jupiter's globe if the "seeing" is bad. The best all-around nights for observing are those with a combination of both good transparency and good "seeing."

Is there anything favorable to be said for cloudy nights if you want to observe astronomical objects? Your only compensation may be a lovely one: with the right kind of clouds, you may be treated to the astronomical-meteorological combination of the moon (or sun) producing in those clouds disks of color called cloud-coronas. Or, with other clouds, you may see the

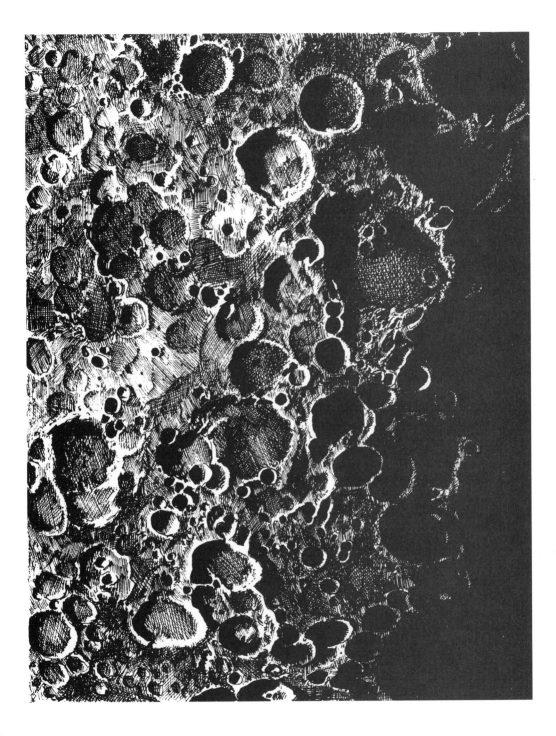

luminous and sometimes colorful arcs, circles, and spots of the various halo phenomena. The most famous halo is the huge "ring around the moon."

The best time to look at a celestial object, or even to look for a celestial object otherwise not seen, can depend on more than surrounding conditions of clear air or still atmosphere or moonless sky. There are many entirely astronomical factors to consider: when the object you want to see is closest, or otherwise best presented in space to our Earth.

Want to see a "shooting star," a meteor? With patient sky watching on a clear, dark night you will see a few. But there are certain nights, the same predictable nights, each year when your number of meteors can be many times better. Those are the nights of the annual "meteor showers," when Earth is passing through a greater concentration than usual of these rocky objects in space. Usually after midnight (it varies according to the specific shower), you may see as many as several scores of meteors per hour on these astronomical holidays.

The best time to see most of the planets is when they are near *opposition* (opposite the sun, thus rising at sunset, visible all night long, and also closest and consequently brightest and biggest). These oppositions occur at some-what different dates each year and must be learned from almanacs or the like if you cannot distinguish the planets from one another (and from bright stars) and thus notice when the planet is beginning to rise near sunset.

A few planets have special conditions of visibility which can help you locate and identify them even if you do not have an astronomy reference work at hand. Imagine a world which sometimes outshines the brightest star and is often closer to us than any other planet but which has been seen (even unknowingly) by only a small percentage of people. It remains unseen even by many amateur sky watchers, and it was never observed (legend says) by the famous astronomer Copernicus! Such a planet exists, and its name is Mercury.

The explanation for Mercury's elusiveness is its closeness to the sun in space and therefore also to the sun in our sky. Mercury whizzes around its small orbit in just eighty-eight days, and the periods when it moves out to fairly good visibility on either side of the sun (above the west horizon after sunset or above the east horizon before sunrise) are only a few weeks long—at most. There are further complicating factors, but for viewers in mid-northern latitudes (like the United States, Canada, and Europe), the angle of Mercury above the sun at dusk is steepest within a month or so of the start of spring. Thus even if you had no other information, you would know that a bright point of light low in the west about thirty to sixty minutes after sunset for a week or two sometime between February and May is probably

Mercury. (The only other object you might occasionally confuse for Mercury in such a situation is Venus—usually much brighter, brighter by far than any other point of light in the heavens.) And if you did not have at least this simple information? You might never notice this fascinating fellow world of Earth, this often orange spark of beauty which I frequently watch to the call of returning whippoorwills in spring.

Some sky phenomena are far more spectacular than Mercury, sometimes covering vast regions of the heavens, but they can be even more elusive because of their time of occurrence and their brevity. The Northern Lights, or aurora borealis, is frequent only above fairly high north latitudes, but wherever you are you increase your chances of seeing it if you keep track of solar activity—this you can do in part by watching sunspots (with proper, safe observing techniques) and being aware of the peaks of solar activity which occur at roughly eleven-year intervals (the next maximum is likely to occur in the early 1990s). Aurorae are most common near maximum solar activity with some of the greatest displays tending to occur a year or two after the peak. The best time of night for the Northern Lights is also important to know. It is usually the middle of the night. People who hear on the news of a potential great display often give up and go to bed too early: a half-hour after they lower their bedroom shades, the previously quiet sky is alive with spectacular, moving arrays of colored patterns.

Another example of awesome sky-spanning beauty which so many people needlessly miss is the rainbow. Rainbows are ephemeral, and your chances of glimpsing one are greatly increased if you know where to watch for their appearance: always in the direction opposite the sun. There are other important pointers for would-be rainbow watchers (the rainbow can appear even if it is not raining right at your location; the rainbow is below the horizon when the sun is more than about halfway up the sky). If you know where a rainbow will appear (if it is going to appear), you can make sure that your view of that sky-region is unobstructed and keep checking right there at frequent intervals as the shower passes and the sunshine strengthens. Adding even a few more rainbows to your life is invaluable.

One of the most remarkable cases of a sky spectacle "hidden in plain sight" is the halo phenomenon called the *circumzenithal arc*. As the name indicates, it forms an arc centered on the zenith (the overhead point of the sky). When the sun gets fairly low and the proper ice crystals exist in the proper clouds, this amazing curve of all the rainbow colors can appear. The trouble is that it appears high above the sun, higher than most people ever look. Again and again—a number of times each year—I have had the strange experience of seeing people who would marvel at a rainbow walking about unknowingly while just above their heads in the sky hung an arc of colors virtually as beautiful. It is shaped like a smile, a smile of nature amused at our unawareness, or a diadem of beauty sitting upon the head of the sky and also crowning our ignorance of its presence or even existence. The observer who sees it feels privileged and proud that he or she is not one of the unknowing throng walking by, but such a person should not be overly proud. Even those of us who know to look high above the sun for it whenever the sun is low and wispy clouds feather that part of the sky, even we forget to look or are sitting inside most of the times it appears. More important, what similar crown or adornment or astonishment of beauty sits over our head or behind our back or beside our feet while we look elsewhere unknowing? By gazing *only* to the sky we may miss the rarest and loveliest wildflower blooming right in our path.

While we sleep some nights, meteors by the score zoom overhead, and Northern Lights dance and burn. Indoors for ten minutes at the key time in twilight, we miss a curtain of significant color draping the entire western sky. Is there any cure for our blindness? If we know that there is a time and place for every thing in the heavens as well as the Earth, and we set out to learn as many of those times and places as we can, and we take the steps to bring ourselves under the heavens then, we are making progress toward the goal of gathering up that lost beauty.

These considerations begin carrying us from secrets of knowledge to secrets of attitude or spirit. Although I place a lot of faith in wheres and whens to get people seeing, there remains for all of us one last problem, maybe only solvable by geese across the stars—I am talking about the contempt of familiarity. You will probably never see enough rainbows or Northern Lights to become tired of them. But will your initial interest in seeing a legendary constellation or staring at the intricate face of Jupiter in the telescope weaken, will these frequently-available wonders fade for you? We all realize (occasionally) that we are taking wonders for granted— even taking for granted the people who are closest and dearest to us. But we do not realize this or do anything about it often enough. Most of the time our unexamined assumption is that we know what there is to know, have enjoyed what there is to enjoy, in these things. About any complex and truly beautiful entity—probably any planet or star, certainly any worthy person—this assumption is utterly false. What can we do to remind our- selves that the assumption is false and keep renewing our sense of wonder?

Surely we can get special help from being there at the times when familiar celestial objects are going to be in an unfamiliar situation or form. The most blunt and effective kind of "hide in plain sight" is that of the daily sun or nightly moon—we know them to be the most awesome of all celestial wonders seen from Earth, yet we can have them right before us so often that we forget to really see and appreciate them and their radiance in the world. We desperately need to rediscover them, and not just once but many times. Seeing them in an unusual situation or guise can help immensely. That is one of the great values of eclipses. Strange to say, whenever you watch an eclipse you cannot help noticing before and after what an altogether marvelous thing the sun or moon is even outside of eclipse. The realization comes as a shock.

Eclipses are not frequent, but there are other opportunities of similar kind. In the spring (for North American and European observers), check in an almanac the exact times of new moon. The sunsets which fall between about fifteen to forty hours after those times are the ones you want to be out at. For on those occasions, though sometimes only in the period from about fifteen to thirty minutes after that sunset, you can spot a marvel. Low in the western sky, you will see the crescent moon so slender it looks like a golden hair, sometimes lacking its ends, sometimes even so soon after new moon that its delicate luminous thread is broken into pieces! The "young moon" is there many a month, but few people even know of its existence, let alone the brief when and low where to look for it. The moon seeming to race through clouds or seeming to rise gigantic is a far more common but also wonder-restoring sight.

So is the setting sun with all of its strange color, its flattening, and its other possible distortions and peculiarities, including one of the most stunning (though fairly infrequent) of sky phenomena: the *green flash*. When you have a chance to see the sun set bright and clear and yellow on a horizon a mile or more away, you should always look for the green flash: the last piece of the setting sun (or first of the rising sun) may suddenly burn intensely green (or even blue or violet) in the form of a giant star. It sounds like a fantasy—just as the broken thread of young moon does—but the young moon and the green flash are actual phenomena which you can quest after and see. And when you do see them (or even now, when you read about them), you should understand better that the "ordinary" sun or moon you thought you knew is not so ordinary after all. They are worlds astoundingly different from ours (as the telescope helps remind us), but even just in our sky you cannot always tell when they will treat you to some strange new manifestation of themselves which your wildest imagination never suspected.

While the green flash and young moon and giant rising moon are all restorers of wonder you should look for at special times and places you must know, it is perhaps the advice or warning about beauties appearing unexpectedly which can help remind the lazy, vain, uninterested part of us that we had better be on guard. And, there is another argument, not as easy to believe but once believed perhaps the strongest of all. This is the argument that not just at certain times are the heavens beautiful but—if you look well enough—at virtually all times.

To true connoisseurs of astronomical beauty, the varieties of weather, site, and state of mind ensure that no two astronomical events or scenes will ever be quite the same. One of the most lovely "conjunctions" I ever saw was through a perfectly positioned "window" in the clouds—a lemon moon, spark of Venus, and bright star Regulus of Leo together before me while a gentle rain began to patter and whisper on fallen leaves all around. What a thrill it is to see a section of heavens brilliant with stars through the drifting fall of snowflakes—their sometimes star-shaped crystals seemingly a powdering on one's upturned face from the stellar multitudes themselves. Lightning with a rainbow or meteor passing through it; moving and moon-pierced wall of fog propped up by the apparently vaster-than-Earth kitchen implement called the Big Dipper; tiny distant birds flying through the narrowest gap separating the moon and a planet while cardinal song rinses the ear and spears the heart in the hour before sunrise—all of these combinations of nature, and countless more of your own, you can find. And often, I must admit, not just your heroic eyes but also your hearing, smell, temperature-sense, and flowing thoughts will be in play to help you appreciate all the wonder of the heavens and Earthly scene around you. That is what happened to me on that Halley morning of the geese.

When I lifted to the starry heavens my eye, what did I find? One of the most strange and wondrous sights I have ever beheld.

There is no way the illumination could have come from one country street light, so it must have been from starlight which made them visible as gray objects, maybe twenty or more of them, no brighter than but very much bigger than bright stars. The geese. I was seeing a wide V of dimly, ghostly glowing geese gliding across the stars swiftly. I could even make out the vague outline and movement of their wings. Even though a quite large section of the heavens was visible to me between the surrounding trees, the line of geese swept across it in no more than about two full seconds.

It would be very difficult to portray visually what I saw in those miraculous seconds. Our illustration at the start of this chapter shows something perhaps equally marvelous, the view held by one of the geese well back in one of the arms as its companions precede it seemingly to the heavens themselves. But to my wonderstruck mind and heart a different imagination naturally came: one inevitably thinks and feels the constellations—especially these mighty ones the geese flew across—as huge—huge almost beyond measure. Many of the stars of Orion, and certainly those of the great clusters of the Pleiades and Hyades, really are related and share at least roughly the same area of space, the true expanse across which they are spread being dozens or, in the case of Orion, hundreds of light-years wide and deep and high. So for this line to stretch from the western shoulder of Orion all the way to the Pleiades quite automatically gave my mind the astounding impression that I was seeing a V of geese—or *some* kind of marvelous, solar system–dwarfing creatures—whose line extended for a thousand light-years or more. One of them could have nested in the lovely Pleiades cluster, the line have moved even great Orion the Hunter's strong stance a bit with force and wonder, the flock have reached the edge of our Milky Way galaxy in the minutes that were all that was left to the full darkness of that tiny night of mine on the eastern edge of North America on our little planet. To what distant wintering ground were these cosmic geese migrating? Even superclusters of galaxies hundreds of millions of light-years away would have been a modest, insufficiently far-flung destination for them. Maybe in a few months they would pass the 15-or-more-billion-light-years-from-Earth border of our universe and, having fed and rested first, cross a gulf somewhat larger than that of Mexico—to universes other than our own.

Even in just the two seconds I saw them, I noted the moment when they passed over Halley's Comet's position in the sky and had just an instant enough to expect and then see fulfilled the passage of the V's point right by that of the V-shaped Hyades cluster which outlines Taurus the Bull's mighty face—a larger V by far than even the Hyades (the largest I had ever expected to see in my life), V to V, arrowhead to arrowhead—and one wondering watcher, me, pierced by that double arrowhead so deeply he stood there after the geese were gone for several minutes, helpless with wonder . . . or perhaps helped beyond our capacity to fully understand with wonder, the wonder of that rare vision.

Countless wonders—even ones so great as that of the geese traveling among the stars—are waiting for you to experience. They will continue to wait—that is, to be available—for the millions and billions of years of Earth and the starry universe. But *you* must not wait. Your life may be a thousand cycles of the moon, maybe 10,000 clear and starry nights long, but not likely longer in this our home in the universe. It does not take a heroic effort to let loose that hero of heroes called the eye. But in a world where so many of us, lost in apathy and laziness, deflected by dejection, stare blankly or not at all—a common danger of the human condition—there must be from you that slight-in-effort but monumental-in-effect nudge of yourself to do just one key thing: to look. That is the ultimate secret of seeing.

Once you do look, adventure is unending. You can either not see or you cannot stop seeing (and enjoying and adventuring). Take your choice: in a sense (the sense of vision) there is no in-between. Astronomy may not quite begin with the eye, but that is where you meet or front the starry task and reward with your champion that vision is. The eye is where you enter the starry room of the universe. All the eye can see is not all there is, but it is a myriad of times and things more than you had dreamed. It is an everlasting, ever-growing enough.

And a line of geese crossing the stars.

2

Year of the Mandala and Monarch Moon Eclipses

IN ONE'S PERSONAL LIFE, a certain few years stand out as more profound than all others—years of special excitement and happening that do more than the rest in shaping a person into what he or she is.

The same can be said for years in the lives of sky watchers and their lifelong love affair with the heavens. For me, 1982 is possibly the year that comes first to mind. The most spectacular meteor I have ever seen, my best views of the Northern Lights, all the planets visible at the same moment, the most colorful and enduring twilights of the century: all these, and many more things, I witnessed that year. But somehow, among all these wonders, when I think of 1982, I think—if not necessarily best, yet first—of the two great lunar eclipses. Somehow they seemed, and still seem, the foundation events upon which others were built or around which others were weaved.

There is no such thing as a usual lunar eclipse, or one that transpires quite as expected. Each one is a fresh wonder, and why that is so is especially well demonstrated by the 1982 eclipses. In a world where the Earth's shadow never paints the moon it touches with quite the same pattern of hues or in anything less than beautiful and mysterious ones, there can be no disappointment at the future eclipses we will see. The 1982 total eclipses of the moon are merely outstanding examples of the rarity and distinctiveness that makes all total lunar eclipses memorable. Yet they *are* examples, more moving than theory alone, casting their dark beauty- and meaning-stained light across the years—and, I hope, through these pages.

What were these eclipses? One was the longest in the history of the United States and perhaps the most astonishingly variegated with its shifting patterns of black and deep red, and its brilliant breakout of the moon from the shadow in the end. The other was, by bizarre coincidence, similar in rare appearance to an eclipse that had taken place exactly nineteen years (to the day and hour!) earlier. Both the December 30, 1963, and December 30, 1982, eclipses were added to the rare previous eleven in history that turned not the customary lunar-eclipse orange or red, but black. Virtually all black. Black enough for the full moon to be utterly lost from view by many of its naked-eye observers. And for me, unbelievably, that already strange second eclipse of 1982 had an even rarer and more special ending.

". . . At last will put to flight/A darkness deeper than the night." Those words (or something like them) have long been in my head, waiting for a theme to fulfill, a poem to someday fit. I have not yet found the right place for them in a poem, but the phrase "a darkness deeper than the night" has several times occurred to me when I have seen those marvels of nature called eclipses. Total eclipses of the sun—causing the sudden fall of darkness in broad daylight—are most awesome of all. But, as the summer of 1982 approached, I knew that the total lunar eclipses to come in the next six months were likely to be very unusually thrilling ones indeed.

The planet Earth was being treated to three total eclipses of the moon in 1982. Three times in the course of one Earth orbit the full moon would not pass a little north or south of the Earth's shadow, as it does most months. Instead, it would go right through—and not just through the Earth's outer or peripheral shadow, called the *penumbra,* or partly through the central shadow, called the *umbra.* The moon would pass completely within the umbra three times, for three total lunar eclipses. That is not only the most possible in a single calendar year, it is a total that had not happened since 1917 nor would again for more than 500 years after 1982—in A.D. 2485!

Of course, it would not be everywhere on Earth that got to see three total lunar eclipses in this year, even if there were no such things as clouds. A total eclipse of the moon is visible from only slightly more than one-half of the Earth—the half experiencing night during the time of the eclipse. If you trace out the hemispheres experiencing night during the three total lunar eclipses of 1982 you find that *no* single strip of territory was thrice-overlapped and could have gotten a view of all three events.

I would be content with two total lunars (people in Europe would get just one—the first one that year). Where I live in the eastern United States, people will not have a chance to see two in one year again until 1996. And as the summer of 1982 approached I was especially starved for sight of Earth's shadow totally covering our one fair, bright, large natural satellite.

Why? No total lunar eclipses had been visible from New Jersey since 1975, when two were seen despite some problems with clouds and twilight.

TOTAL LUNAR ECLIPSES 1982-1999

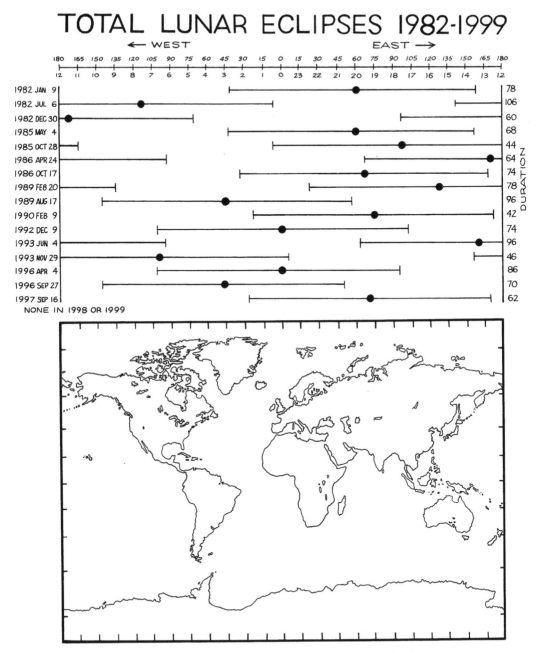

Dates, locations, and durations (in minutes) of total lunar eclipses, 1982–99.

I had been unable to see the moon become totally eclipsed just as it set and the sun rose on September 6, 1979—for a very unusual reason. My part of New Jersey was then under a tornado watch because of Tropical Storm David's center passing to our west. What an eerie feeling it had been to lie in bed with open windows, forty- to fifty-mile-per-hour gusty and fitful winds, and sporadic but sometimes heavy rain in the dark of 4:30 A.M.—especially eerie because I had known that a tornado might be forming somewhere near in the night and that the moon was entering Earth's shadow beyond the vast structure of wind and rain and cloud, beyond the carefully watched eye of the storm.

In June 1982 I was also thinking ahead to how starved I might become for eclipses after that year. After 1982 no more total lunars would be properly visible from New Jersey until 1989. Unless I traveled very far, the 1982 events would be my only opportunities for seven years either way in time—the fourteen years of my young adulthood.

Millions of other people were in the same boat. To see the shattering beauty of a total solar eclipse you almost certainly have to travel, for any given spot where you might live will experience one on an average of just once every few centuries. Yet somewhere in the world there is a total solar in most years. In 1986 and 1987 there were only marginally total solar eclipses, and there are none in 1989, 1993, 1996, and 2000. But often several years may pass without any total lunar anywhere in the world: 1987–88, 1994–95, and 1998–99 are examples of such periods. (Incidentally, there was no total eclipse of the sun in 1982, but *four* partial solar eclipses—so the year had the maximum number of eclipses possible, seven.)

These statistics about the frequency of eclipses have very personal consequences for those of us who watch the heavens. Total lunar eclipses are rare enough to occur only a small number of times in the life of an alert skywatcher, and one tends to recall all the personal aspects of those times by the vivid aid of these slow, rich, magical events—so good for and encouraging of pondering (of all kinds). A man or woman has got to be a stone not to be moved a bit by the sight. But you also find yourself reflecting on what life is like for you in that season and year as you gaze for an hour or more at the familiar lunar face altered—seemingly itself dimmed and flushed with feeling thought.

Frequency was important, but I was also interested in another statistic about the eclipse of July 6, 1982. And that was its length.

The eclipse would feature the longest totality (period of total eclipse) visible from the United States since before the country's founding—since 1736! The last one visible anywhere on Earth that was longer was in 1859.

An old (but usually reliable) book I checked said the longest lunar totality possible was 103 minutes; this totality was to be just over 106 minutes.

A more intensive study of eclipse statistics turns up facts that lessen the impact of these statements, of this title of the Longest American Eclipse—but only a little. Total lunar eclipses in 1953, 1964, and 1971 were all about 100 minutes long. On July 16, 2000, there will be another one of 106 minutes—about as long as the July 6, 1982, eclipse. (Then there will be none of more than 102 minutes even through the year 2050.) Furthermore, when I write here of the longest eclipses I am referring only to the period of totally eclipsed moon. A lunar eclipse is quite interesting, though, from the moment the umbra first touches the moon until the moment its last edge moves off the moon. In this measure the eclipse in 2000 is the longest between 1950 and 2050—3 hours and 56 minutes, or about 2 minutes more than the figure for July 6, 1982.

Another measure of eclipses produces a rating of them in order similar to that by duration, but not quite the same. For the moon to be within the umbra a long time it must take the longest path through the circular projection in the sky of that cone of shadow in space. The longest path is straight through the center of the umbra. We can get an idea of how far in the moon gets by what is called the "magnitude" of the eclipse (this should not be confused with the magnitude one meets more commonly in astronomy, a measure of brightness). If the farthest the moon gets into the umbra is one moon-diameter, then its magnitude is unity, 1. That is the smallest magnitude a lunar eclipse can have and be total. An eclipse with a magnitude of 0.50 in this system would be a partial one, in which at best only half the moon's diameter is covered by that central shadow of umbra. This does not mean, by the way, that 0.50 or 50 percent of the moon's *area* is covered.

What is the greatest magnitude a lunar eclipse can have? According to the astronomical calculator Jean Meeus, the largest in the twentieth century is not the 1982 eclipse (with magnitude in umbra of 1.717—the moon 1.717 of its diameter within the umbra's edge) but that of the July 26, 1953, eclipse, with its magnitude of 1.864. Not until November 4, 2264, will an eclipse have a larger magnitude (1.869). How can the 1953 eclipse be shorter than the 1982 one and yet have a greater magnitude in umbra? The catch is that the apparent diameter and apparent speed of both the moon and the umbra at the moon's distance vary considerably. It all depends on the moon's distance from Earth and the Earth's distance from the sun at the time of the eclipse. Calculating how long the Moon will be totally eclipsed is not a simple matter.

●　　●　　●

As I prepared for the Long Eclipse of 1982, one other factor led me to expect that it would be unusual. The great length I could count on, but the color and brightness of the eclipse were another matter. No one could predict exactly what they would be, but a certain event and the consequences I was already seeing made me suspect that the color and darkness would be of the most unusual kind. The event was the tremendous eruption of the El Chichón volcano of Mexico that past spring. The consequences of the eruption—which I was already seeing, thousands of miles away from the volcano several months later—were the most intensely colored and lasting twilights of the century.

I started seeing the dramatic twilights in New Jersey in early June. I will not comment on them in great detail here, but I will say that they were caused by a haze of high-altitude sulfuric acid (yes, sulfuric acid!) that formed (and, amazingly, kept forming) as water vapor combined with immense quantities of sulfur dioxide released by the volcano. This sulfur dioxide was drifting round and round the world at latitudes farther and farther from that of the volcano. The sulfuric-acid haze was at such high altitudes that it remained lit—gloriously lit!—by the sun for a long time after sunset occurred for viewers on the Earth's surface. And that haze was surely going to redden and darken some of the sunlight bent 'round the Earth's solid body into the Earth's shadow.

What does the bending or refracting is the Earth's atmosphere. It is the secret of a lunar eclipse's color. If Earth had no atmosphere, our planet's shadow would be entirely dark. Someone on the moon would simply see the sun blotted out by the dark hulk of Earth. Instead, in most total lunar eclipses, ordinary sunlight shining on the moon is replaced by the light from a ring of red silhouetting the giant Earth in the moon's sky. And we on Earth see not a black moon but an orange or rosy or coral moon. That color is a sort of ultimate extension of twilight: every place along the ring of sunset and sunrise that encircles Earth contributes its reddened light to the Earth's umbra (which is thus more a beam of color than of darkness).

The color on the eclipsed moon is thus not just beautiful to see but also marvelous to understand. And its understanding is all the more involved and marvelous because the color is as changeable as the atmosphere that helps produce it. If an eclipse occurs when a high proportion of Earth's twilight zone is cloudy, the eclipse will be somewhat darker—at least if the part of the umbra that the moon passes through corresponds to an area of the twilight band that is cloudy. But ordinary clouds are generally confined to only the lowest five or ten miles of Earth's atmosphere. The atmosphere higher than that is rather clear and permits light to pass through without

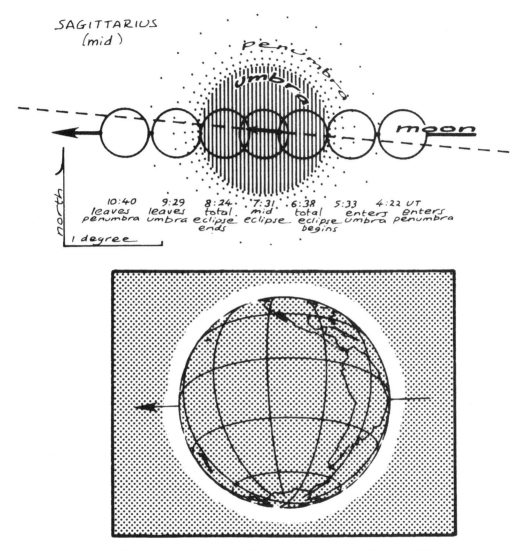

SAGITTARIUS
(mid)

penumbra

umbra

moon

north

10:40   9:29   8:24   7:31   6:38   5:33   4:22 UT
leaves   leaves   total   mid   total   enters   enters
penumbra   umbra   eclipse   eclipse   eclipse   umbra   penumbra
                           ends   begins

1 degree

Moon's path through umbra, lands facing Moon, at eclipse of July 6, 1982.
By Guy Ottewell.

too much reddening or dimming, so a lunar eclipse is never really black—unless a major eruption of ash or sulfur dioxide has recently occurred.

Those clouds of ash or hazes of acid have now been implicated in all the dozen cases between 1600 and 1981 in which a total lunar eclipse was so dark that the moon disappeared or almost disappeared from naked-eye view. Was the thirteenth going to occur on July 6, 1982? The El Chichón cloud was already proving itself remarkably dense and wide-spreading. And yet I wondered. By early June it had reached only my home latitude of 40°N. in its northward movement, and only about the equator in its southward spiraling around the world from Mexico. The moon on July 6 would be heading not through the north part or south part of Earth's shadow but right through the middle—corresponding roughly to Earth's equatorial regions. Independent of other factors, the moon is somewhat darker when it passes through the center of the umbra anyway. But whether the El Chichón cloud had extended far enough into Earth's equatorial regions to much further darken the eclipse remained to be seen.

I was ready. I would rate the eclipse's darkness on the customary Danjon scale (see accompanying table). I would also try taking off my rather strong glasses and comparing the brightness of the out-of-focus eclipsed moon with that of out-of-focus stars whose brightness was known to me. I did not even know for sure in late June whether I would be doing this rating, and enjoying the Long Eclipse, from my home in New Jersey. As it turned out, the spot I observed the great eclipse from was not far from the geographical center of the North American continent, a good deal farther north and more than a thousand miles west of New Jersey—in North Dakota. It also turned out that my very high expectations for the eclipse were not going to be disappointed. Rather, they would be exceeded. I would find myself being overwhelmed by a truly haunting sight, by a beauty more than I had bargained for: a moon part deep-red, part black; a moon twilight-stained, volcano-struck, hours Earth-shadowed, ever-changing through creeping moment after moment.

Danjon's Brightness Scale for Total Lunar Eclipses
(L stands for Luminosity.)

L = 0   very dark eclipse; moon hardly visible, especially near mid-totality

L = 1   dark eclipse; gray to brown coloring; details on the disk hardly discernible

L = 2   dark red or rust-colored eclipse with dark area in the center of
the shadow, the edge brighter

L = 3   brick-red eclipse, the shadow often bordered with a brighter
yellow edge

L = 4   orange or copper-colored, very bright eclipse with bluish bright
edge

When I arrived at my friend's place in North Dakota, I discovered that
something had just beaten me there: the El Chichón cloud. That first clear
night an intense and prolonged twilight greeted me. My friend said that
such twilights had begun just a few days before. The deep-red twilight might
set the tone—quite literally—for the eclipse.

A few miles northeast of the little town of Parshall, a group of us set
up early on the big night. Actually, at this northerly latitude just a few
weeks after the summer solstice, true night did not come for quite a long
time. Here at the extreme western edge of the Central Time Zone and the
northern edge of the contiguous United States, we were experiencing just
about the latest sunset and latest end of evening twilight ever possible from
anywhere in those forty-eight states: the sun went down at 9:50 P.M. CDT,
and twilight lasted until at least 12:13 A.M. CDT. The longest eclipse I had
ever experienced was going to last for virtually all of the shortest night I
had ever experienced. As a matter of fact, the first edge of the Earth's
outer, lighter shadow, the penumbra, touched the moon at 11:22 P.M. The
penumbra is generally too light a stain to be seen until it covers at least
half of the moon, though. Twilight would not really bother the period during
which any part of the Earth's shadow on the moon would be visible from
our site. If I had stayed in New Jersey, twilight would have interfered
somewhat with the end of total eclipse. Where in the world would be the
best place to see this eclipse—in the sense of having the moon right overhead
at mid-eclipse and at midnight? Somewhere in perhaps the most deserted
part of the South Pacific (Easter Island would get a very good view).

But neither twilight nor the relative lowness of the moon in our sky
promised to be our problem that night. Nor were the mosquitoes, which
at first were as bad as South Jersey's (which is saying much!) but soon were
set off us by the arising of a bit of breeze. No, our problem was one that
New Jersey was not having that night, though my chances to avoid it had
been better here than back home: clouds. They were troubling the moon,
and more threatened. How much of the Long Eclipse would we see?

We saw a corona, a blue-and-red circle around the moon caused by
cloud droplets. Then, for a while high cirrus ice-clouds formed dimly but

definitely the upper half of a vast (22° radius) halo—the famous "ring around the moon." A rather prominent *parselene* (mock moon or moondog) also appeared briefly. And there was a distinct display of actual "moonbeams" scattered up from the beclouded lunar disk. Of these M. Minnaert wrote: "This very rare phenomenon conveys an impression of ominous gloom." We did not think of them as a bad omen for our chances of seeing the eclipse. We were still hopeful that the clouds would break in time, and those strange moonbeams merely added to an enjoyable mood of foreboding.

At last we could all see the penumbra plainly on the left side of the moon. Right on schedule, at 12:33 A.M., twenty minutes after the official end of astronomical twilight, the umbra appeared and began its dark creep across the moon—and dark it did look, even in the six-inch telescope. We guessed that this was going to be the dim eclipse anticipated.

But now, after we had watched the moon pass through a set of "phases" such as it never normally has and noted the curved edge of shadow that visually demonstrates our Earth is round, the clouds came back and we lost sight of the moon. Worry increased as more and more minutes passed. As the beginning of the total eclipse neared, however, most of the sky over us had at last cleared, and we were amazed and gladdened by the astonishing wealth of stars that appeared now that the full moon was almost completely shadowed. Then, just after the start of totality, the moon itself broke out into the clear for a few minutes.

We were truly shocked. The moon was not completely dark; all of it had not disappeared from the effect of volcanic haze in our atmosphere, for the haze had not yet spread much to the south of the equator or to northern latitudes a lot higher than ours. And so it was only most of the middle and top half of the moon that were truly black to the naked eye as the total eclipse began. A huge slice of the south part of the moon was a deep orange-red or red, and there was a slight rim of light left on the moon's northern edge. Never had I seen so variegated a moon-appearance. We wanted 106 more minutes of this! But it was not to be. The clouds swept back in to frustrate us, swallowing up the astounding vision. Would we get the next large break in time?

We did. Mid-totality was at 2:31 A.M. CDT—only one minute before the precise minute of full moon, an indication of how nearly central the eclipse was. Just before then, a phantom moon broke out. I was awed by the beauty and strangeness of the scene. The moon as a whole was dimmer now. Earlier, the bright parts had amounted to a brightness similar to that of the star Altair—about magnitude 0.8. Now the parts with any brightness left at all combined for a magnitude of only about half of that—about magnitude 1.5, or slightly fainter. More than twenty stars in the heavens are brighter than this, with the North Star a little dimmer. But the eclipse light was spread out over a large part of the moon, not in a point like a star, so to the naked eye this moon seemed to be some grim piece or dollop of dim, deep-red planet or comet or nebula. Or it was a giant ember almost dying (imagine our moon burned out and fading forever!), glowing feebly but with eerie potency there in the south. A marvelous contrast to the sight of this dim red ember was the spectacularly bright, white star clouds of the summer Milky Way, the best of which hung in awesome stillness just to the right of the moon there in Sagittarius, there in the haunted prairie dark.

In the six-inch telescope, the whole moon now seemed a gray ghost, only just barely luminous. No lunar features, not even the seas, were distinctly visible at this point. I gazed in deep admiration at this view of the dark but full and big ball silhouetted against space and a dusting of stars. Faint stars were visible right up to the moon, and I saw a very faint one pass right behind the edge of the moon for an "occultation."

About fifteen to twenty minutes after mid-eclipse, though the moon was still deep within the umbra, astonishing brightness changes began to occur from minute to minute, and a brightening slice of moon was widening and shifting position. The lunar seas could now be made out through the telescope. Obviously this part of the Earth's shadow was not so strongly darkened by the volcanic haze.

The end of the long totality was in itself probably the most spectacular I have ever seen in a lunar eclipse. The sliver of moon escaping from the umbra was incredibly sharp and bright, and its quickly widening yellow contrasted gorgeously with the still red-and-black rest of the moon. This was yet another face of the great eclipse, like some strange, precise, and multicolored symbol in the sky. The earlier two-toned moon had been pied; now we were seeing a giant mandala, that mysterious circular symbol of some Eastern philosophies which contains cryptic interior patterning.

Clouds were at last coming back, the long morning twilight was just becoming distinctly visible, and we were about to pack the cars and prepare to leave while glancing at the final part of the continuing show. But then someone shouted, "Look in the north!" Suddenly there had sprung up behind us, atypically only near the end of night, a fairly dim but quite distinct group of long, vertical green rays—the Northern Lights! The aurora was paying a tribute to the eclipse in its movements. It was a fittingly rousing and grand end.

● ● ●

Normally, you give one Danjon rating for an eclipse—or, in unusual varying eclipses, one at a time for different brightnesses at different parts of the period of totality. The July 6, 1982, eclipse refused to be treated so simply. Most observers agreed with my friend and me that the pied moon during much of that totality deserved a Danjon rating of L = 0 in some patches and L = 1 in others (with a few slighly brighter splotches?). Some observers argued for L = 2 for much of the moon during much of the eclipse, perhaps on the basis of the L = 2 description's referring to "dark red or rust-colored" and the L = 1 description's mentioning "gray to brown coloring." It is true that the Danjon description of a dark area in the center of the umbra with edge brighter could be regarded as an approximation (very rough indeed) of the situation in that night's umbra—but it could also apply to somewhat brighter or darker eclipses.

What then could be concluded? The Danjon descriptions are inevitably somewhat imprecise as to color. Their remarks about visibility of the moon and of lunar features are clear—and certainly are well in accord with L = 0 and L = 1 for the July 6 eclipse. But we must not expect too much precision in verbal descriptions of colors even while we know it would be a loss to not have them at all as guides. Visibility of features and brightness are more accurate criteria. To judge from observers' opinions in eclipses of recent decades, an eclipse of L = 3 might have a total brightness of magnitude − 2 or a lot brighter, and L = 2 might be around magnitude 0. But what would be the magnitude of the very darkest lunar eclipse?

That question was no longer a theoretical one as I pondered photographs, people's verbal descriptions, and my memories of the July 6 eclipse throughout that autumn. The July 6 eclipse with its L = 1 patches that El Chichón had missed was not that darkest eclipse. It just missed qualifying for the list of only a dozen eclipses known during which the moon virtually disappeared, during which the Danjon rating would be L = 0 or *less* (note that the Danjon description for L = 0 does not quite say "invisible"; it says "hardly visible, especially near mid-totality"). But the question of how dark the darkest eclipse would be was an active one in my mind because the upcoming eclipse of December 30, 1982, just might possibly turn out to be that.

As John and Stephanie Mood point out (in a superb article on El Chichón and the dark 1982 eclipses in the February 1985 issue of *The Griffith Observer*), the popular astronomy magazines did not yet seem to understand the nature of the El Chichón–dark eclipse connection—the magazines were suggesting that the December 30 eclipse ought to be lighter because the moon was passing not centrally through the umbra but rather high through the northern part of it. The only question in my mind was how far north the El Chichón haze would spread and how much it would thin out by December. Would the darkening effect of its covering a larger area offset the lightening cause by reduction of the cloud's density?

The answer was a shock. The El Chichón cloud not only spread farther—over almost all the world—by December, it also became its densest over most parts of that spread *in* December! How could this be? At the time, I could not figure it out. I heard reports about the cloud's spreading to farther latitudes, and I could see for myself that in New Jersey, October, November, and especially December had the strongest twilights and dimming of sun and day sky yet (the strongest these things were ever to get). You see, at the time I and everyone (save for a few researchers waking to a great discovery) were assuming that El Chichón's cloud was composed of fine volcanic ash. There was no way that such a cloud could spread farther and yet get denser. But the process of the invisible sulfur dioxide gas combining with water vapor to form sulfuric acid could take place more slowly and reach a maximum density of the acid haze much later—in December 1982.

Still ignorant of this surprising process, I was nevertheless expecting a very dark eclipse. I scanned over the list in Edward Brooks' fascinating article in the June 1964 issue of *Sky & Telescope* (a list adapted from a 1963 article by Frantisek Link): the $L = 0$ eclipses. Before 1601 we do not have accurate enough descriptions to identify $L = 0$ eclipses with much certainty. But in that year, Kepler was one of the observers and describers of the very dark eclipse of December 9. This was the first entry on the list. But I had seen for myself—as a child—the final eclipse, the one to which Brooks was devoting the article. Or, rather, I must say that this was the eclipse which I had *not seen,* during totality—the nine-year-old me knew enough about lunar eclipses (having even seen one before) to fill up with a delicious dread at the fact that the moon at that dead-of-night and dead-of-the-winter time had simply disappeared in a brightly starred sky. A lunar eclipse was, I knew, not supposed to happen that way! But it had, and many observers had even rated this eclipse as below zero on the Danjon scale. The figure obtained by taking the average of all the good ratings sent to *Sky & Telescope* turned out to be $L = 0.2$. What was the brightness in terms of magnitude? There was an estimate of about magnitude 4.1. In other words, the brightness of the entire moon was no greater than that of a single star of less-than-average naked-eye brilliance. So little light spread out over so much area made the moon at that eclipse invisible to most viewers.

Dark Lunar Eclipses (L = 0)

Lunar eclipse	*Associated Volcanic Eruption*
December 9, 1601	Huaina-Putina, Peru, 1600 (uncertain)
	Kamchatka Peninsula, Russia, 1600 (uncertain)
June 15, 1620	Mt. Hekla, Iceland, 1619
December 9, 1620	Mt. Hekla, Iceland, 1619
April 14, 1642	Mt. Awu, Indonesia, January 1641
May 18, 1761	Jorullo, Mexico, Sept.-Dec. 1759
June 16, 1816	Tambora, Indonesia, April 1815
October 4, 1884	Krakatau, Indonesia, August 1883
October 16, 1902	Mt. Pelée, Lesser Antilles, May 1902
	St. Vincent, Grenadines, May 1902
April 11, 1903	Mt. Pelée, Lesser Antilles, May 1902
	St. Vincent, Grenadines, May 1902
	Mt. Santa Maria, Guatemala, October 1902
March 22, 1913	Mt. Katmai, Alaska, June-Oct. 1912
September 15, 1913	Mt. Katmai, Alaska, June-Oct. 1912
December 30, 1963	Mt. Agung, Indonesia, March 1963
December 30, 1982	El Chichón, Mexico, March-Apr. 1982

Now it was natural that I should think back to this dark eclipse as I prepared for what I suspected would be another dark one. But there was another link (and not Frantisek) between the two eclipses, a very odd one indeed. It was the timing of the events.

First of all, the dates were the same: the earlier eclipse had taken place on December 30, 1963, nineteen years to the day before the eclipse I now anticipated. This coincidence is not unique to those particular full moons. The period of 19 years—19 orbits of Earth about the sun—is almost exactly equal to 235 *lunations* (a lunation is the period from one occurrence of a lunar phase—say full moon—to the next, about 29½ days). Whatever the phase of the moon tonight, that will be its phase exactly 19 years from tonight—sometimes give or take a day. This is known as the *Metonic cycle*.

But the Metonic cycle does not guarantee that two eclipses nineteen years apart will take place at exactly the same *time* of day—even though the same time is necessary to having those eclipses identically visible from the same places around the world. The second timing coincidence of the December 1963 and December 1982 eclipses was that their time of day *was* almost the same—the first one had its mid-totality at 6:08 A.M. EST; the second one at 6:29 A.M. EST. Their duration and magnitude and path through the umbra were not quite the same—the earlier eclipse was a little longer, a little closer to the umbra's center, and it passed south (not north) of the center. But the moon's position in the heavens was almost identical—a point near the feet of Gemini the Twins. More significantly, the whole backgrounds of the experiences for an observer were almost identical: you would get up at the same hour, look in the same place, and if you lived in New Jersey have morning twilight interfere with the latter stages of both eclipses.

There was thirdly and finally one other truly amazing coincidence concerning the timing of the eclipses, and it was the most significant. In both cases, the eclipse was preceded almost exactly nine months by one of those rare eruptions that can fill Earth's atmosphere with incredibly high-altitude haze for so long. In March 1963 the eruption had been that of Mt. Agung in Indonesia, fifty years after the two L = 0 lunar eclipses of 1913 affected by the 1912 eruption of Alaska's Mt. Katmai. Some of the most explosive and ashy eruptions of the twentieth century had had far less profound and lasting effects on Earth's atmosphere and twilights (Mt. St. Helens is a prime example). But Katmai and Agung did not do as much as El Chichón. One had to go back ninety-nine years to find an eruption that did. That was the one that had darkened the lunar eclipse of October 4, 1884. It had come from a volcano called Krakatau—more popularly known as Krakatoa, whose 1883 eruption produced what was probably the loudest explosion on Earth in thousands of years.

The day and evening before the eclipse of December 30, 1982, was a frantic scramble for me, given all the things I had to do. I was fortunate to receive the generous loan of a few extra tripods for telescopes and cameras. But all this preparation looked as if it might well be wasted: even a few hours after sunset on December 29, the sky was completely overcast, with a little fog and a light rain. Then I heard a very favorable weather report. Could I trust it? Would the sky clear in time?

By 8 P.M., the moon was becoming dimly visible behind thick clouds. By 9 or 10 o'clock it was showing more clearly, but still plenty of clouds were present. By 11 P.M., light clouds were racing with breathtaking speed over an otherwise clear full moon that kept tinting them—by diffraction caused by their droplets—with lovely tones of azure and pink. At 11:30 P.M., the highest full moon of the year hung near its maximum altitude in the south, filling with its brilliant radiance the entire boundless cup of an absolutely clear sky!

After getting very little sleep, I rose at 3 A.M. and drove with a friend and a carload of equipment to meet another friend at our observation site,

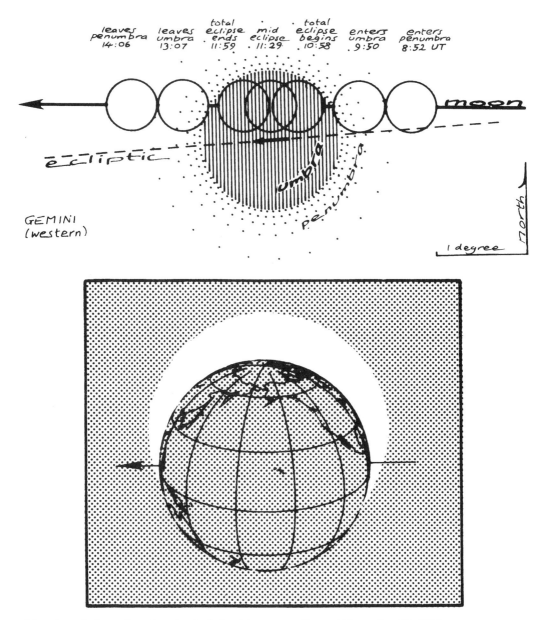

leaves penumbra 14:06 · leaves umbra 13:07 · total eclipse ends 11:59 · mid eclipse 11:29 · total eclipse begins 10:58 · enters umbra 9:50 · enters penumbra 8:52 UT

moon

ecliptic

umbra

penumbra

GEMINI (western)

north

1 degree

Moon's path through umbra, lands facing Moon, at eclipse of December 30, 1982. By Guy Ottewell.

36

East Point (which actually gives one a view west—where the moon would be—over Delaware Bay). On our drive down, bright moonlight kept flashing between trees as if from frames of a moon-made movie as the car raced on to a meeting with the moon—and the quarter-million-mile extension of the night-and-twilight called the Earth's umbra.

But our first meeting was a surprising one with the penumbra. The first edge of the moon would enter the outermost penumbra at about 3:52 A.M. EST. But the general rule, remember, is that the penumbra must be across about one-half of the moon before any of it becomes visually detectable as even a slight stain. And the penumbra would not be halfway across until roughly 4:20 A.M. We did not have to be set up until around then, I thought. Thus it was very much to my surprise when we arrived at East Point at 4:07 A.M. and found that the penumbra was already quite distinctly visible! The penumbra is often so light and elusive that beginners are never sure they are seeing it at any stage. But by 4:20 A.M. on this day the penumbra was so dark it must have tricked many people into thinking it was the umbra. What, then, would the umbra itself look like?

The answer came on time—at about 4:50 A.M.—when we could see the first tiny black bite of umbra almost immediately even with the naked eye. With a breeze off Delaware Bay and a temperature in the mid-20s, it was chilling out (I later developed a horrendous head cold), but not so chilling as the beauty of that dark and so unusually sharp-edged shadow.

As we watched the detailed ball of the moon through the 8-inch reflector, $4\frac{1}{4}$-inch richfield telescope, $3\frac{1}{2}$-inch richest field refractor, 8×50 binoculars, and other instruments, we could see the exquisite chiselry of lunar maria, mountains, craters, and other features being overwhelmed utterly by the vast wave of inky darkness. When Tycho (the moon's most prominent crater) was reached, its brilliance (the diamond pendant of the Lady in the Moon) was blacked out without a glimmer by the hypnotizingly slow and seemingly heavy yet gliding edge of umbra.

At one point, however, I noticed through the telescope a beautiful, transparent band of slightly red light bordering the outer edge of the umbra as it advanced across the lunar surface. I do not know the explanation for this effect, but it lasted for only a few minutes at some time around 5:05 A.M. The time was (again roughly) about 5:20 A.M. when we briefly glimpsed with our naked eyes the very feeblest glow of deep red in part of the umbra. These two fleeting events were the only traces of red we ever saw at any stage of the eclipse!

As the remaining slice of moon got still more slender, I noticed in the telescope that the horns or cusps of this crescent were of different lengths, one becoming so far extended that it was a very strange deformation of the umbra's edge. By this time, what before had been a sky of only a few stars drowning in a sea of moonlight was now a dark, dark winter heaven, with stars sprinkled everywhere and several fine planets parading with bright stars in the southeast.

Most striking was the arch of brilliant winter stars that now shone prominently around the faltering lunar phantasm in the west. The stars Castor and Pollux, heads of the Gemini twins, were not too far above the almost-eaten moon. Through the scopes, numerous little stars already pricked the sky all about the almost wholly gray moon-ball—some perching right on the edge of the ball, which appeared very three-dimensional in its silhouetting against them and the sky.

When would the last tiny piece of unshadowed (or un-umbraed) moon be extinguished? Would the moon be visible at all to the naked eye in total eclipse? How would this totality compare with the one of exactly nineteen years before? The answers were threatened by cloud banks low in west and east, and by morning twilight not far away.

A few hours before the chilly break of the next-to-last day of 1982, the full moon had been brightly silvering the waters off East Point, where the (un-bothersome—because red) light of a lighthouse was blinking behind us—me, two friends, four telescopes. But that moonlight, which had filled the sky's bowl with its milk, had been soon touched with a stain as deep as blackest dried blood. Now the stars around the moon were burning ever brighter, seemingly in amazement at the fading specter that consisted of only a narrow yellow sliver on the very dimmest gray ball.

Through the optical instruments there was one last show before totality. Both of the cusps of the yellow crescent—the longer in the north and shorter in the south—were now blue. The blue horns did not lose any of their length as the yellow between them shortened and shortened, finally becoming no longer than they. Did I miss what happened next? I do not think so: my eyes were almost constantly at one of the telescopes. I think that the colors merely faded into a whitish light in the course of just a couple of minutes.

And then it was 5:58 A.M.—start of totality! But there was still a strip of light at the moon's upper right edge. It almost seemed bright, almost like a part of the uneclipsed moon—but this was of course only in contrast to the rest of the moon. It was a comparatively brighter area within but near the edge of the umbra. Soon it also became obvious that the entire lower left quadrant of the moon was a bit less dark than the rest (excepting, of course, the much brighter upper right strip). The bright strip, which crept counterclockwise along the northern rim of the moon, was caused by the part of the umbra farthest from the umbra's center. It was not really bright (no color) and had certainly not seemed so when the leading and upper edge of the moon had passed through this same portion of umbra before totality. Clearly, much of what is seen and not seen in lunar eclipses is a result of contrast—especially between light and dark, or fairly dark and very dark. Another example is the fact that just before totality much of the moon's left edge (already well within the umbra) was not distinguishable from the sky even in the eight-inch telescope but that soon after (when the distracting light of moon outside of umbra was gone) the left edge was again quite detectable.

It was now well past 6 A.M., and the first traces of morning twilight were growing stronger in the east. What did the moon I was seeing the areas of dark and more dark on with the telescopes look like to the naked eye? With the naked eye, even before any twilight interference, I could not see most of the moon at all. Those parts I did see must have been the brighter strip and the less dark lower portion: the eclipsed moon looked like a big, upside-down, and very dim comma! When I looked back through the telescope I saw a faint star seemingly perched right on the Moon's edge.

Not only twilight but those clouds still low in the east and west were threatening an all-too-early end to the adventure. Surely the moon would be lost in those clouds at the western horizon before it started coming back out of totality just before it set and the sun rose. I told myself I had been lucky to see any of the eclipse at all, let alone the many dark wonders of it I did. After all, my weather log had already recorded sixteen overcast days in that month of December, and two more—the last two days of both the month and the year—were on the way. The eight hours or so of clear sky with the eclipse in it were the only such period in what proved to be eight straight essentially overcast days from December 26 through January 2. I had already seen enough of the eclipse to know that some people might lean toward an L = 1 (at least for parts of it) but that most would rate it (as I did, even before mid-totality) an L = 0. Early in totality I had compared the giant comma out of focus with stars out of focus (my glasses were off) and thought it not much fainter than magnitude 1.5 Castor. But I did not doubt that that magnitude would drop considerably by mid-eclipse. And I was right: the entire moon disappeared or was practically invisible for observers farther west in the United States without trouble from twilight. At that point some people rated the magnitude as about 3.0, others somewhat fainter. The conclusion was that this eclipse was extremely dark, though not quite as dark as that of December 30, 1963. Thus it joined the ranks of the very few known eclipses of L = 0 that have occurred since the year 1601.

Mid-totality was at 6:29 A.M., but our sky was getting too bright to allow a view of the moon, even if it had been in a somewhat less dark eclipse. One more glance at the clouds below where the lost moon was convinced me that there was no hope of seeing it start to come out of the umbra as it would just before moonset. The eclipse had been a wonderful experience, but all of us were cold and tired, and it would take many minutes to get all the equipment back into the car.

My reasoning was not bad, but I am sure that it was really exhaustion (I had slept only a few minutes that night) that almost made me miss the remaining—and climactic—sight that I will never forget. We were actually driving away, me in the lead, down the long road back from East Point to Heislerville, when I took a last, tired look across the flat marshes to the west. And there it was: a slender, upside-down crescent of deep-orange moon hanging just a few degrees over Delaware Bay!

The clouds had broken up in that spot. The orange was the first returning slip of moon outside the umbra, orange only because it was so low in the sky. The crescent was bowed away from the horizon—upside-down—as one would never, never see otherwise. I pulled the car off to the side of the road and, my eyes fixed on the spectacle, pointed in silent awe at it for my friends driving up behind.

In those last electric seconds my wits did not quite fail me. Racing back on the empty road through the marshes, I came to the place where East Point Lighthouse and the moon appeared side by side, abreast of each other; flung the doors of the car open in mid-road; flew my fingers across camera and tripod and telescope attachments—all while absorbing this astounding sight. A tiny, thinner-than-the-thin-moon strand of distant geese, whose voices came to me in the growing light for just a second, trickled in the violet west toward and then, just barely missing, past the sliver moon and the still calmly, slowly red-blinking lighthouse. The descending band of violet was Earth's shadow projected on our atmosphere to the west—and then onward that quarter-million miles to where the moon was just starting to edge out of it.

The impossible upside-down crescent was visibly sinking, now lodged between the lighthouse's base and the very cedar trees that a dear friend and I had seen and touched when their branches were literally covered with hundreds upon hundreds of migrating monarch butterflies on the warm, sleepy afternoon of October 9. Now the last and impossibly vast and slow butterfly of orange moon was departing. But not before I had gotten a few fragile photographs and, far more importantly and preciously, had seen it and remembered it—perhaps the last such sight of it to be possible at the lighthouse for a hundred years or very much longer. I do hope that fine old lighthouse is still there to greet the next monarch moon when that moon passes East Point on its migratory way from eclipse at a far-future dawn.

After it sank below the cedar trees, I managed to catch last sight of the moon as it faded out just before touching the level waters of the bay. It was about two minutes later that the sun, which had risen behind a narrow cloud bank while I last watched Luna, finally burst out in a brief break to purple the dark clouds and make glisten the stalks and brown heads of tall wind-nodded reeds. And so at a day just begun, in a year almost ended, lighthouse and reeds, moon and sun, geese and remembered monarchs, and the very volcano-sooted black shadow of the planet we live on, had all met to make joy in a most tired, scrambling, and utterly fortunate watcher.

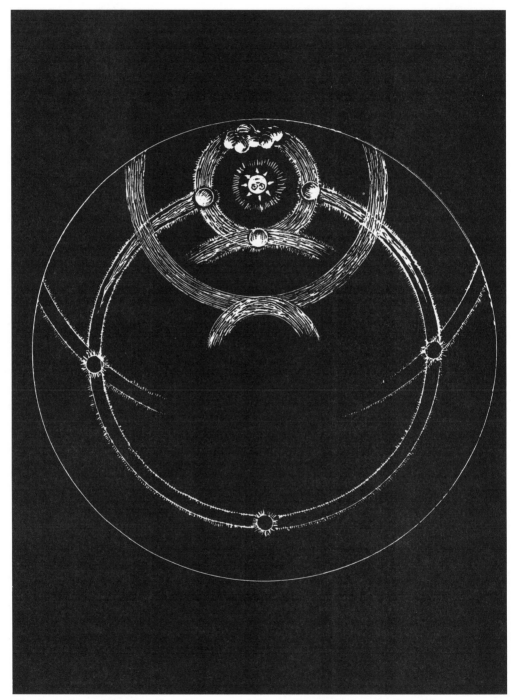

The Seven Suns of Hevelius (based on a more elaborate drawing by Hevelius himself)

3

The Many Suns of the Daytime Sky

THERE IS ONE TRUE SUN easily visible in the daytime sky. Like all great things, however, that Sun has its imitators. Not many people have seen these sham Sols, but that is mostly a measure of how seldom people look up. Even many amateur astronomers and weather watchers who do gaze heavenward frequently have not seen the other suns because they tend to look in well-known times and places for things that are written up a lot in books. The other suns are not mentioned much in books, and admittedly some of them are rare. But at least one of them is extremely common, practically weekly, occasionally occurring off and on for much of a day. All you need to know is where and when to expect these phenomena.

They are not really astronomical objects, of course, though they are caused in part by the true Sun, which is one. All but one of these kinds of fake suns are also caused by certain kinds of ice crystals and belong to the rich and varied family of "halo phenomena." Most halo phenomena are in the form of rings or arcs, but the ones that appear as spots or patches of light have been compared to the sun (sometimes actually mistaken for a cloud-veiled Sun!) and bear this in their names. Some are always colorless, others may share beautifully in the sun's color, still others can display the full spectrum of rainbow colors.

If one sun is not enough, prepare to meet more! Included in our survey are the mock suns, the subsun, the anthelion, the paranthelia, the double sun, the almost unbelievable countersun, and, last but very certainly not least, the Seven Suns of Hevelius, of which six were imposters.

● ● ●

Mock suns are as common as any halo phenomenon and can be seen at least dozens of times a year in most climates. They are remarkable in being sometimes extremely bright and colorful. They are often seen with the 22° halo, which most people best know as "the ring around the moon," but which is more vivid (or at least more colorful) when produced by the sun. This halo has a radius of 22° (about two fist widths at arm's length), and the mock suns are roughly elliptical or irregular-shaped patches of light located just a little outside this halo, but always at the same altitude as the true Sun. The mock suns appear farther out the higher the true Sun is, but they seldom occur at fairly high solar elevations and cannot appear at all when the Sun is more than 61° above the horizon.

Mock suns may show no color at all but usually show at least some, occasionally all the colors of the rainbow. The hues most often seen (in my opinion) are red, yellow, and blue (in order of distance from the true sun). Mock suns can be blindingly bright, casting shadows, and it is possible that very rarely they may even produce secondary mock suns—mock suns of mock suns! Mock suns can appear without the 22° halo visible, or with it; in a pair (one to either side of the sun), or alone.

Scientists believe that mock suns are caused by squat hexagonal plate crystals of ice that have their broad sides oriented close to the horizontal. Such crystals occur in cirriform clouds, so whenever the feathery cirrus clouds drift in their delicate threads and wisps around the sun's part of the sky, look for the sudden staining of the clouds with light and color in the mock sun positions (except, of course, when the sun is very high). Scientists often call mock suns *parhelia* (the singular is *parhelion,* which is Greek for "with the sun"). Farmers and other outdoor people who notice them usually call them *sun dogs.* And indeed mock suns occasionally have a long horizontal ray of white or bluish-white extending outward (away from the true Sun) from them, which is called a "tail"! Once in a while you may even notice that there is a dim ring of light all the way around the sky at the Sun's altitude, often called the *parhelic circle* (because the parhelia, as well as the true sun, lie on it).

A halo phenomenon that is supposed to be quite commonly visible from planes is another one of our solar impostors, the subsun. It appears at the same azimuth (compass point) as the true Sun, but always at a lower altitude—in fact, an altitude below the observer's horizon (thus visible only in ice clouds below you). It is usually a rather long ellipse (one of the most common kinds of UFO as such!) but becomes rounder the higher the true sun is above the horizon. The subsun is actually an image of the sun reflected up from almost horizontal mirrors of ice crystals floating in clouds below. It is the extreme far-from-the-sun form of a *sun pillar,* which is a column of light that can appear sticking straight up from the true Sun or straight down from it, usually near sunset.

The brilliance of the strange subsun can be so great that it may produce a 22° halo if its own as well as other halo effects. At least one authority, Robert Greenler, believes, however, that "subsun dogs" (which seem to be mock suns caused by the subsun) are not really caused by the subsun.

Unlike mock suns, the subsun is not formed by reflection and refraction, just by reflection, and it is therefore itself colorless.

One of the least documented but perhaps not the least rare of halo phenomena is probably nothing more than a certain kind of incomplete sun pillar. This is the startling double sun, which occurs when the true Sun seems to have a duplicate of itself hanging just a degree or two above it or, very rarely, below it! M. Minnaert states that there have been just a few cases in which there were two or even more of these apparent solar images visible at the same time as the true Sun they mimic. He offers the explanation that they are really bright detached pieces of a very prominent sun pillar. I agree with this theory because in a number of sightings I know of (including ones of my own), there was a more or less obvious bar of less-high-altitude cloud above the sun with the double (fake sun) just over the bar. Yet the almost perfectly spherical shape and the sun-like size of the doubles I have seen have been amazing—they did not *look* as though they were a section of sun pillar. And there was no obstructing bar of cloud visible to cut the supposed pillar off sharply at the top edge of the double.

The double of the Sun can be very nearly as bright as the original and takes on the same color (as do sun pillars). Keep an eye out for this spectacular sight; I suspect that it is far more common than has been believed.

A very uncommon halo phenomenon is the anthelion, whose name means "against the sun," in the sense of opposite it (in the sky). But the anthelion is not (in all but the unusual case of sunrise or sunset) located at the "anti-solar point" (around which "the glory" and, at greater angular distance, the rainbow both arc). The anthelion is opposite the Sun only in azimuth. Its altitude is the same as the Sun's. The anthelion appears as a colorless bead, usually with the also colorless parhelic circle on which it lies. In addition to the Sun, parhelia (mock suns), and anthelion on this circle, there may also be two other uncolored spots of light at 60° azimuth to either side of the anthelion—the paranthelia. The paranthelia may be visible more often than the anthelion, but none of them are likely to be noticed by any but the most dedicated halo seekers except when they occur together as part of great complex displays when the entire sky is covered with cirrus clouds of proper thickness and proper ice-crystal orientations. Very rare and bizarre arcs and pillars of various kinds have been observed with these fake suns so far away from the true Sun. Experts have attempted to explain by the use of computer simulations what crystal orientations would produce the various effects. Interestingly, there do not seem to be reports of the anthelion when the true Sun is more than 46° above the horizon, and in one computer model this would indeed be the limiting altitude.

Perhaps even more startling than any of the phenomena yet discussed is what Minnaert called the countersun. This is the image of the Sun on water that seems to rise out of the lake or sea to meet and melt into the setting true sun! The phenomenon is also observed in reverse at sunrise. The question is: How is it possible?

The countersun is an "inferior mirage" ("inferior" only in the sense that it occurs *under* the real Sun). It occurs when there is a sharp temperature gradient in the atmosphere near the surface and therefore a sharp density gradient. The strip of light showing the true horizon is refracted (bent) so strongly upward that the light passes well over an observer's head, reducing the apparent angular distance between the true Sun and its reflection on the water. Most often the countersun seems to raise only a small part of itself before meeting the true Sun. But I have seen as much as *half* the countersun get above the horizon before the two suns touched and began to melt into each other through an ever-widening "waist" of light. The countersun is something you must see to believe—and even then you may find yourself wondering if you are dreaming in another world, so astonishing is the sight. The countersun image is so good that you might even see sunspots on it—though they would be upside-down because the image is inverted. Strange and beautiful distortions of the setting sun occur at most clear sunsets, but the drama is increased by the difficulty of knowing just which ones you will see that particular night. Might this be the night you see the wondrous "green flash" or the awesome countersun?

Last of all, as proof of how many "suns" may be found in the daytime sky, we come to the Seven Suns of Hevelius. Johannes Hevel or Hoevelcke, best known as Hevelius, was a seventeenth-century astronomer who lived in Danzig, which is called Gdansk in Polish. Among his many accomplishments were the discovery of four comets, the charting of the moon, the measuring of the stars' positions, and the invention of six of the eighty-eight constellations now considered official. He also discovered the duplicity of several of the most famous double stars and built some of the longest refracting telescopes ever made. But it was on February 20, 1661 that Hevelius made perhaps his most unusual observation.

That was the day he saw the great halo complex. His original drawing shows a number of halo phenomena now well known to experts. There is the 22° halo and the 46° halo (partly cut off by the horizon since the Sun's angular altitude at the time was roughly 26°). We also see on the sketch the upper tangential arc to the 22° halo, the circumzenithal arc, and the parhelic circle. Of the seven "suns," some are also easy to account for. There are the true Sun, two mock suns, and the anthelion. The sun drawn where the upper tangential arc touches the top of the 22° halo is unidentified but presumably not a discrete effect, just a temporary local concentration of light due to the cloud distribution—one would expect the glow to be brightest at this point of intersection. So five of the suns are explained. It is the final two, and the arcs on which they lie, that still have experts wondering.

Those final suns are not paranthelia. They are located on the parhelic circle at 90° of azimuth from the true Sun (Hevelius saw suns at the same altitude at four evenly spaced points of the compass!). They are located at the intersection of the parhelic circle with two immense arcs that look as if they are parts of the most giant halo of them all. The 22° radius halo impresses people with its immensity, and the 46° radius halo (though seldom if ever seen complete) has its bottom just below the horizon when its top is at the zenith. But Hevel's halo (as it is often called) must have a radius of somewhere between about 90° and 98°—perhaps too large to fit in its entirety even in the whole dome of the sky! There have been a few reports of what were probably sections of Hevel's halo in the 300 years since his sighting, and possibly of his final two suns—which have been called "90° parhelia" for lack of a better term (the paranthelia are sometimes called "120° parhelia," lying as they do 60° from the anthelic point and 120° from the true Sun). The problem is that so few people know anything about halo phenomena. Such effects as Hevel's halo must be rare, but if there were more people knowledgeable about the common halo phenomena we would surely get a few observations—and photographs—that would suffice to solve the mystery of what ice crystal arrangements cause it and when to expect it.

The special phenomena of "meteorological optics" are ones that every lover of the sky should learn about. Not just rainbows of all kinds, but cloud-coronas, glories, dozens of different halo effects, the intricacies of the blue sky, and numerous low-sun phenomena like the green flash and countersun await. The fellowship of people who know when and where to look for these marvels is a tiny one, but it need not be—and should not be if we are going to learn more about their mysteries. More than elaborate scientific equipment, what is required are alert and knowledgeable amateur sky watchers all over the world.

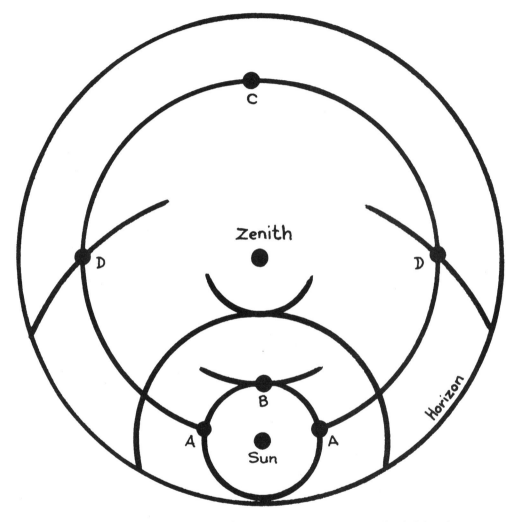

The Seven Suns of Hevelius. Sketch identifying "suns": A—mock suns; B—brightening on upper tangential arc; C—anthelion; D—90° mock suns of Hevel's Halo.

I hope that touring the topic of the many "suns" you never knew existed has stirred your desire to learn more about such phenomena and to start looking. Just imagine discovering a new halo arc or spot and having that delicate, rare and mysterious bit of glowing, ice-projected sky-geometry bear your name—like the Parry arcs, the circumzenithal arc of Bravais, Bavendick's Ellendale display, and others. Imagine yourself making a drawing of a sky full of miraculous patterns and labeling it, as Hevelius did on the top of his, with the words *Septem Soles:* Seven Suns!

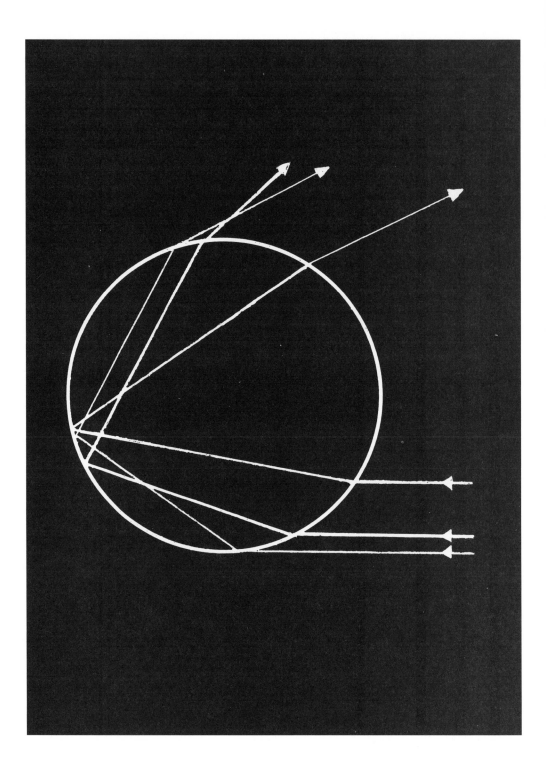

4

100 Rainbows

As I RIDE past the wide lawns on this street, the sun is getting low at the end of a clear day. Sprinklers are watering those lawns. And in every one of them is a rainbow. Many of the fountains and waterfalls of the world have the correct size of spray droplets and correct angle on sunlight to produce a rainbow—countless thousands of them shining at this instant, right *now* as you read. But only for that he or she who stands in the right place to have sunlight not just reflected back from the drops but also refracted into the full range of sunlight's component colors at the angle of 42° (four fist-widths held out at arm's length) from the *anti-solar point* (which lies exactly opposite the sun, at the shadow of the observer's head).

Yet it is not usually just one place at which these conditions are met in an actual rain shower. For every section of road, wood, or city for miles— and then different miles as the shower moves on—there is a different rainbow. Even you and I, standing side by side, see ever-so-slightly different rainbows if the droplets producing them are fairly far away, radically different bows if the droplets are close (as in a fountain or sprinkler).

Even each eye can see its own and different rainbow.

And each droplet itself bears its own personal rainbow if the sun or other suitable light source is shining upon it.

The world is a-tangle with countless rainbows yet, of course, seeing any one of them is not a common experience—especially seeing the traditional giant one of the sky, of the receding storm, which remains even the most experienced rainbow seeker's favorite. And always the strange nature of the rainbow adds to the sense that, if in some ways omnipresent, the rainbow is also rare. And unique. Does anything else present a sight so prominent, so seemingly touching the world, yet so impossible to reach—not even an object but instead what is to be seen in a very specific set of directions relative to sun and rain? How odd that a set of directions relative to sun and rain should be so beautiful, and this beauty both most peaceful and most stirring at once! How appropriate that the thing which may be the most *purely* lovely is the most unvanquishable and the most elusive, even harder to catch or catch up to than any *rara avis* forever fluttering beyond reach! Few people would argue that a thing whose reflection may remain visible while it disappears (see rainbow #67 in the pages that follow) is not the most elusive in the world! But is the rainbow the most purely lovely? If the emphasis is on purely (in the sense of exclusively), I think it must be, for the rainbow is nothing *but* beautiful and what pertains to the beautiful.

Is the rainbow the most unvanquishable of things? It is at least as unvanquishable as many other things, and it is, in a good sense the word no longer has, more *flagrant* (that is, literally, flaming into notice) than any of the others in its undefeatability. If man ever destroys all rain and splashing or springing water, all places where rainbows hide ready to leap out, or if he ever "succeeds" (for such success is ultimate failure) in hiding the sun (that true natural—though not spiritual—lord and father of the rainbow as the rain is the true lady and mother), then the last man or woman will die as the seemingly last rainbow fades from sight. And after humans are gone at last, even should it take 10,000 years to clear, the smoke *will* clear and from clouds struggling to be clean the first beam of sun will burst through to paint again, first shyly, then like the banner of an eternal victory— whether mankind has abandoned it (and died) or rushed to its open arms to embrace it (and lived)—the rainbow.

I began with an invocation (of sorts) to the multitude, actually the manifold infinities of rainbows in the world. I moved on to invoke *the* rainbow—the epitome of beauty, peace, elusiveness, longing, and mystery which with its single and singular glory spans the eternal sky and our spirits in a vision lying beyond (or through or on) every individual manifestation of the phenomenon. But for most of this chapter I wish to move from the spiritual and aesthetic (if that is indeed possible) to the practical—yet by no means mundane. The title of this chapter, you see, is not a reference to one hundred "little" rainbows in one hundred water sprinklers on a millionaire's estate. For, like any good rainbowologist (if there were any such thing), I am going to name and number for you as many different kinds of rainbows as I can—in this case, one hundred different kinds. Is it possible to name so many? Will I have to cheat to do it? Well, three of my "rainbows," though looking somewhat like rainbows and commonly mistaken for rainbows, are not really rainbows. I believe they deserve inclusion here, though: think how marvelous something must be to be *mistaken for a rainbow!* What about my other ninety-seven varieties of the rainbow or rainbow phenomenon? I think you may be surprised to see how well I do!

1. *Primary rainbow.* The one that most of us have seen—and would be lucky to see, even partially and fleetingly, one hundred times in a life (if you do see this many, every one will be different enough to be distinguishable from the other ninety-nine if you make paintings of them all).

2. *Secondary rainbow.* It is not really rare, just uncommon to see a full, strong example of this dimmer but roughly twice-as-wide rainbow. Look for it outside of (above) the usual primary rainbow, at about 51° from the anti-solar point. The order of colors is reversed from those in the primary: the red in the primary is on top, the red in the secondary on bottom—so they face each other. The secondary rainbow is caused by a second reflection of light within raindrops before it emerges.

3. *Primary and secondary rainbows with Alexander's dark band.* Light reflected once within raindrops escapes at no angle greater (measured from the anti-solar point) than that of the primary bow, but (to some extent) at all angles smaller (in other words, all the sky below or inside the curve of the bow). Light reflected twice within raindrops escapes at no angle smaller than that of the secondary bow but (to some extent) at all angles greater. In short, there is added illumination in the sky from these reflections inside the primary and outside the secondary but not between the primary and secondary. In most cases you can easily see that the sky is darker between the two—as was first recorded by Alexander of Aphrodisias early in the Christian era.

4. *Double rainbow or swimmer's rainbow or rainbow of each eye.* You could call a display of primary and secondary rainbows together a "double rainbow." But that name could also go to the side-by-side intersecting arches that swimmers can sometimes behold in spray that they splash up so close to themselves that each of their eyes sees a different rainbow.

5. *The tertiary or third-order rainbow.* This rainbow, caused by three reflections in the raindrop and very much dimmer than the secondary, has been seen on rare occasions in nature. It is wider still than the secondary (each higher order rainbow is wider than the one before it). By the way, all the world was looking in the wrong direction for this bow until Edmund Halley calculated correctly that it occurs back in the direction of the sun.

6. *The fourth-order rainbow.* Never seen in nature, but spotted in laboratory (and home!) experiments.

7. *The fifth-order rainbow.* Observed in nature by the nineteenth-cen-

tury scientist Eleuthère Mascart. This rainbow is about 7° wide (almost five times as wide as the primary) and occurs inside Alexander's dark band with its red partly overlapping that of the secondary bow.

8–21. *The observed higher-order bows from sixth-order through nineteenth-order.* None of these has been observed in nature, but in 1868 Felix Billet observed all the first nineteen orders in a thin stream of water produced by a device he had made. (This is a big help in my quest to list one hundred rainbows, but a legitimate one—as the next entry is not!)

22. *The rose of rainbows.* Not really a rainbow but an enchanting name that Billet gave to his diagram showing the positions and relative sizes of the rainbows of the nineteen orders. If you consider the sky's (never-perceived but existent) display of these nineteen (and more) orders as the "rose of rainbows," then perhaps this entry really is legitimate.

23. *Jearl Walker's dozen rainbows in a drop of water.* If you want to look for the rainbows of the first twelve orders or more for yourself at home, read Walker's "Amateur Scientist" column in the July 1977 issue of *Scientific American.* (For Jearl fans: the column does not fail to mention Walker's legendary and rhetorically invaluable grandmother.) The fact that these are rainbows you can see in a single drop makes them worth an entry. After all, in a way this drop or droplet is a kind of rainbow "atom"—the smallest unit of matter that can produce a rainbow, not the element "rain" but the atom "raindrop."

24. *Farthest rainbow.* A friend of mine observed a section of rainbow in a shower which the weather radar he worked at showed must be about eighty miles away. Greater distances would certainly be possible (maybe especially from an airplane). In one way this is not just the opposite of the nearest rainbow but also of the single raindrop rainbow. Or rather it is a rain shower being reduced in apparent size as much as possible by distance— the smallest span or arc of rainbow per rain shower possible as opposed to the largest extent of rainbow (full rainbow) per amount of rain (or tap water). This kind of rainbow would probably be the one in which to see the (angularly) shortest section of rainbow. It would also be a way to see the next rainbow.

25. *Rainbow through haze.* This is frankly not a very desirable rainbow, because the haze dulls not just the colors coming back to the observer but also (unless all the haze is low-lying) the sunlight getting to the raindrops in the first place. I have seen for myself, however, that such a bow has its own interest and identity if not great beauty. An eighty-mile-distant rainbow would have to be somewhat (but apparently not always fatally) dulled by the great journey of air and by the dust in the air through which its light must travel.

26. *Closest rainbow.* For a true natural rainbow—bow caused in falling rain—a record that M. Minnaert mentions is one seen in front of a wood that was just three yards away (and thus the rainbow formed in rain closer than three yards!). As you will see, fogbows and mistbows and tropical storm rainbows of a certain kind can produce even closer natural bows!

27. *Rainbow with one end closer than the other.* This is common because raindrops causing the bow you see may not be arranged in a band perpendicular to your line of sight. Speaking of the ends of the rainbow—it is only the fact that there is usually not enough rain between you and the ground that cuts off a rainbow at the horizon.

28. *360° (full-circle) rainbow.* That may not be so as seen from an airplane. From that vantage point it is possible, but apparently very rare, to behold a full 360° circle of rainbow. I actually have some cause to give a separate entry for rainbows with one or more ends extending in front of landscape as seen from ground level. I recall a striking example in which a friend and I were chasing the ideal rainbow-producing section of a storm in the Midwest and suddenly saw a rainbow with its prominent end passing in front of the bright green field that inclined slightly up before us less than one hundred yards ahead. What must people who like the end of the rainbow think when they find that the full—360°—development of the rainbow is endless? There is no pot of gold at rainbow's end. But every rainbow is a treasure better than gold.

29. *Rainbow with middle closer to viewer than ends.* Must be a bit more unusual than #27, but quite possible.

30. *Two-level deep rainbow.* If two separate rain showers were both causing a section of rainbow in the same direction but (of course) at different distances from an observer, the rainbows would be of equal width and would coincide exactly in the position in which they were seen in the sky. But the particular prominence of the various colors (depending in large part upon drop size) could be different. It might be possible to distinguish when you were seeing such a two-level deep rainbow. Such a happening would be rare because the clouds of the first shower would be likely to block sun from the second unless the latter was much farther, and in most lands so perfectly placed a succession of abrupt showers and sun would not occur frequently. Less rare but still interesting would be different parts of what seemed like one rainbow in the sky caused by showers at very different distances. It would be strange to see one end of a rainbow caused by a shower a half mile away and the other end by a storm fifty miles distant. Both ends would be of the same apparent width: all primary rainbows are of effectively the same angular width, whether in a storm far off or in sprinkler spray twenty feet distant.

31. *Red rainbow at sunset (or sunrise) or during twilight.* As the sun sinks, the colors begin disappearing until only red is left. The reason is that the sunlight is becoming increasingly red (because of scattering or absorption of the other, shorter wavelengths—which are the other colors in sunlight—in the trip through the longest possible path of atmosphere when the sun is low). The red rainbow may be seen until well after sunset, because at high altitudes the rain is still being lit by sunlight even quite a while after an observer at ground level has seen the sun disappear. The ends of this rainbow fade out until only the middle (contributed by higher drops) is visible, then a single red spot, then nothing.

32. *The highest rainbow.* The previous rainbow is the primary rainbow highest in the sky in angular measure that one can see. Its top is more than 42° above the horizon—almost halfway up the sky, with as much as a half-circle of rainbow above the horizon in the sky.

33. *Rising rainbow.* When the sun sinks to 42° above the horizon, the top of the rainbow can (with rain in the right place) be seen rising on the opposite horizon. This is also the *lowest* rainbow except for looking down on various kinds of bows which can be seen well below the horizon (in dew, fog, waterfall, and so on).

34. *Setting rainbow.* When the sun rises to 42° above the horizon, a rainbow's last, top edge sets at the opposite horizon.

35. *Morning rainbow.* For meteorological reasons, this is much rarer than an afternoon rainbow.

36. *Rainbow preceding a storm.* This is even more unusual but quite possible (I know someone who has seen it). Normally, of course, the rainbow is the pledge of peace at storm's end. In fact, it has been argued from Biblical tradition that the rainbow must have been the last of all types of things created—for we read that God first put it in the sky after the Flood as a sign to Noah and the passengers of the Ark that God would never again overwhelm the world with water.

37. *Rainbow with lightning striking through it.* I have seen this well in the Midwest United States.

38. *Rainbow cut off by falling snow.* Snow occurs high up in thunderstorms and looks like downward-hanging tufts if seen falling (in summer never to reach the ground unmelted) from the clouds. I have seen rainbows cut off by this, because, of course, the rainbow phenomenon cannot occur in snow crystals.

39. *Rainbow over a snowy landscape.* Quite possible if you get rain after snow—but rainbow expert Robert Greenler took a now-famous trick photograph of a rainbow over a snowy landscape. The rainbow was not caused

by rain; it was caused by spray from Niagara Falls (out of view in the photo) in winter.

40. *Rainbow with a tornado.* Tornadoes often occur in the southwest and thus right rear quadrant of certain thunderstorms. So it is not rare to see one receding in the late afternoon with bright sun shining on it and on heavy rain near it. Travis Tull has photographed an instance of a rainbow with a tornado—nature's most peace-inspiring (yet stirring) sky phenomenon together with nature's most violent and terrifying sky phenomenon.

41. *Rainbow with supernumerary arcs.* Before I continue with any further rare rainbows, here is a quite common one—yet one that most people do not comprehend. If you see extra bands of pink, green, or white inside (that is, below) the violet of the primary rainbow, you are seeing supernumerary arcs. *Supernumerary* means above or beyond the ordinary number—in this case, the ordinary number of bands in the rainbow. These arcs are caused by *diffraction* and are thus akin to the special colors in clouds around the moon and the beautiful colors in oil splotches on the road.

Very rarely, supernumerary arcs may be seen also to the outside of the secondary rainbow. I could count this as a separate entry here but will not.

42–47. *Primary rainbows with different raindrop size causing different prominence of colors.* These six are really just different major types of the primary. The classification is derived from Minnaert:

Raindrop diameter, 1 to 2 mm. Rainbow with very bright violet and vivid green, pure red, scarcely any blue. Pink and green supernumerary arcs are numerous (as many as five) and merge into the primary bow but begin to overlap and become difficult to distinguish if raindrops are much more than 1 mm wide.

0.50 mm. Red weaker in primary bow. Fewer supernumeraries, but they are still pink and green.

0.20 to 0.30 mm. No more red, but rest of bow broad and well developed. Supernumeraries pale and yellowish.

Less than 0.20 mm. A gap between primary bow and the first supernumerary arc.

0.08 to 0.10 mm. Bow broad and pale, only violet vivid. First supernumerary well separated from primary and is whitish.

0.06 mm. Primary rainbow contains a distinct white stripe.

Of course, you will see all these features (such as the supernumeraries) only if you have very good conditions (sufficient rain and sun) with these various raindrop sizes. But what happens to the rainbow if you get plenty of sun and droplets less than 0.05 mm?

48. *Mistbow.* If rain droplets get smaller than 0.05 mm, they are considered mist. The bow caused by mist has lost its colors and is therefore sometimes called the "white rainbow." But that name is also given to a more commonly seen example of the rainbow phenomenon occurring in extremely tiny droplets.

49. *Fogbow.* A fogbow can be glimpsed surprisingly often in the fog with a dark background and a streetlight behind you—though they also appear beautifully when the early-morning sun shines on fog. Since fog can be very thick, it is possible to see such a bow not only as a giant arch on the road ahead of you (with street-light behind you) but actually, in good cases, as a 360° fogbow. You can even have its lower arc run across your body if you lie down! You then have a mystical bow seeming to touch you. The 360° fogbow is sometimes called Ulloa's Ring. Fogbows can be about twice as wide as rainbows, can have a little blue on the outside edge and orange on the inner, and can be of variable—but substantially less than 42°—angular distance from the anti-solar (or anti-streetlight) point.

50. *Fogbow with supernumeraries.* Fogbows have strong supernumeraries, and instead of the first being pink and the second green, (as with the rainbow), one finds the opposite order.

51. Secondary *fogbow.* This is quite rare, but I have seen it.

52. *Cloudbow.* Fog is a special kind of cloud on or near the ground, but essentially the same phenomenon can be seen high in the clouds from an airplane (or mountain). Some of the special vantage points possible on cloudbows (especially for viewers in airplanes) are different, though.

53. *Tropical storm downpour rainbow.* I was once walking in torrential rain from a tropical storm (far from being a hurricane) at night and could see a white arch like a fogbow ahead of me when the streetlight was a proper distance behind me. Was it caused by mist coming up from the impact of the rain on the road, or were the droplets blown so hard in the wind that they were broken apart and became smaller than those of ordinary rain?

54. *Searchlight rainbow.* The rainbow caused by a searchlight is especially strange because it moves with the movement of the beam. The rainbow actually is seen to slide up and down the beam and even to disappear briefly. See the article on this in the *American Journal of Physics,* Vol. 43, p. 453 (1975), by J. Harsch and Jearl Walker.

55. *Waterfall rainbow.* Quite understandable (forms in spray from impacting water), but beautiful.

56. *Fountain and lawn sprinkler rainbows.* "Manmade" (though not ultimately), but close and interesting to experiment with.

57. *"Upside-down rainbow."* Not really a rainbow, but the average

person who sees one thinks it is. It is not a full horizon-to-horizon span and is always seen well above the sun. It forms an arc around a point that lies at the zenith and is therefore called the *circumzenithal arc.* Like a rainbow, it can have all the colors of the spectrum and appears only when the sun is fairly low. It is caused by ice crystals and is a type of halo phenomenon. Very lovely!

58. *Straight, ultra-wide rainbow that isn't a rainbow.* This is actually the *circumhorizontal arc*—the complement of the circumzenithal arc. It is visible only far below the sun when the sun is quite high. It is rather rare, but its best displays might be the only color phenomenon in the heavens even more spectacular than the rainbow—much wider, and just possibly even longer in its paralleling of the horizon around a sometimes sizable fraction of the full circle of the sky at that low angular altitude.

59. *Circular rainbow that isn't a rainbow.* Quite a few airplane passengers think they have seen a 360° rainbow when they observe a disk of light consisting of concentric circles of alternating red (or pink) and green (or blue). They are actually seeing a phenomenon known as a "glory," centered right around the anti-solar point (you can often see the shadow of the plane on the clouds causing the glory—like the bull's-eye center of the glory's archery target–like disk of concentric color bands). The glory has more in common with the supernumerary arcs caused by diffraction. It is much smaller (that is, nestled closer around the anti-solar point) than the 360° rainbow. The glory must also be very much more commonly seen than the rainbow in its entirety.

60. *Dewbow caused by sun.* The shape of this bow formed in dew—a "rainbow" on the ground!—is a hyperbola (wide-open curve) with arms pointing away from the observer who is nearest to the middle of the curve. Or rather that is how the dewbow is typically seen not long after dawn. As the sun rises higher, it is finally possible—in very rare instances—to see the dewbow become an ellipse.

61. *Dewbow caused by streetlight straight above one's head.* The "anti-streetlight point" is on the ground at your feet. Where is the dewbow you see? Because of the diverging light rays from the light not all that far up above you, the answer is bizarre. The light source lies within the hyperbola or ellipse, and the observer has seemingly deformed inward the part of the curve nearest to his or her feet—in fact, broken it, so that a separate curving part of the figure extends away to either side of his feet (see illustration on page 60).

62. *Secondary dewbow.* This must be possible and on morning dew lie in a hyperbola that is slightly closer to the observer and more open than the primary dewbow.

Strange dewbow caused by divergent rays from streetlight as seen from viewpoint of woman in diagram. (From the reader's viewpoint the dewbow would not appear in this position.)

63 and 64. *Two kinds of reflected dewbows.* We now get into a number of rainbows whose complications and combinations are truly mind-boggling. For our present purposes I will be as short and simple as possible. Our first opportunity for a headache from figuring but a heartache from strange beauty results from the fact that dew can in some circumstances float on the surface of ponds. When this happens, it is typical to see two colorful bows, neither a secondary dewbow. The smaller, less open hyperbola is a reflected dewbow. To make things still more complicated, this bow may arise from the light being reflected from the pond surface either before or after it goes through the dewdrops. Only if the sun gets fairly high can one begin to tell the two varieties of reflected dewbow apart.

65. *The reflected rainbow.* The reflection of the primary rainbow is very interesting indeed. In water, of course, all colors of reflected objects can seem deeper and richer. But note that the reflection of the rainbow is not displaced toward the observer and away from the horizon the way the reflections of clouds and other objects are. If we recall that a rainbow arises

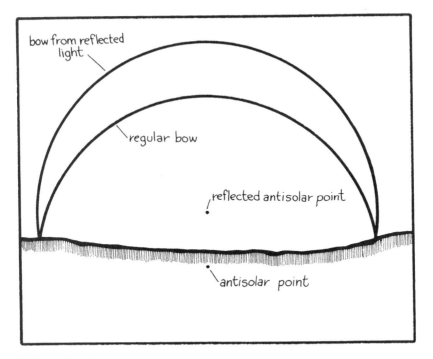

Reflected rainbow.

at a certain angle and is not itself an object in space, it seems to make sense that its reflection would not be shifted, would be perfectly symmetrical with respect to itself in the sky. Thus you can see a rainbow passing in front of a cloud but in the reflected scene on the water appear *below* (closer to the horizon than) the cloud!

But the next set of statements adds a twist that may be even more bizarre than any of the other strangenesses of the rainbow. *What you think is the reflection in water of the rainbow you see in the sky is not. You cannot from your position see the reflection of that rainbow in water. What you see is the reflection of a different rainbow!*

66. *The rainbow without a reflection.* We might almost say that the rainbow does not have a reflection, but that is not quite true. The explanation is this: the reflected image you see on the water is from a rainbow caused by an entirely different set of raindrops from the one you see in the sky. To see the rainbow in the sky causing that reflection you see from the shore, you would have to go out to that point on the water where the reflection is occurring.

Is this exceedingly tricky situation only of technical interest, a mere curiosity? Not necessarily. It would be possible to have considerable differences between the rainbow of the sky and what seemed to be its reflected image on the water below. Imagine, for instance, watching a rainbow in the sky and seeing everything else reflected in a still pond before you except for that rainbow (because the sun and rain conditions are not proper to produce a rainbow as seen from that point on the pond where the reflection would be made). You could even see the rainbow remain while what seemed to be its reflection faded out of view on the water. Or you could behold the eeriest sight of all. . .

67. *A reflection of the rainbow without the rainbow!* The Cheshire cat faded out, leaving only his smile behind. But imagine seeing the rainbow and (what seemed) its reflection on the water, then watching the former fade out while the latter remained.

68. *A rainbow caused by the reflection of the sun in water.* We are not quite done with the various marvels of reflected rainbows yet. Completely different from the previous kind of reflected rainbow is what we could call a "sun-reflection bow." This is a rainbow caused not by the sun directly but by the brilliant reflected image of the sun on water. It is always higher

Rays that cause the rainbow's reflection seen on water.

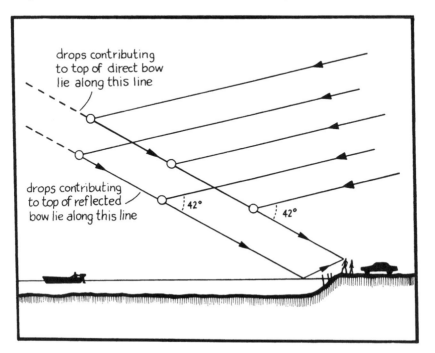

than the ordinary primary bow. If you see the sun-reflection bow or part of it in the sky, you need not be in sight of the body of water causing the reflection. You might behold an "ordinary" reflected rainbow in the water plus both an "ordinary" rainbow and the sun-reflection bow in the sky. Add the secondaries of these bows and you could come up with a truly bewildering pattern of colored circles and arcs spread across water and sky!

A question ready-made to give us headaches: would it be possible to see a reflection of a sun-reflection bow (whether or not it was the reflection of the one simultaneously visible in the sky)?

69. *Rainbow pillar.* Often only a nearly vertical section (one end) of a sun-reflection bow is visible. This is the cause for reports of "rainbow pillars"—except for those cases of people looking toward the sun (the wrong direction for virtually all rainbow phenomena). In those cases of people looking toward the sun, the phenomenon actually observed was probably a vertical section of a halo or that sun-imposter halo phenomenon known variously as a mock sun, sun dog, or parhelion.

70. *Autumn foliage rainbow.* Here is something more understandable and more common: the rainbow seen in front of some especially beautiful landscape. As an example I cite a hill a mile away covered with orange- and red-leaved trees that I saw right through the end of a rainbow. Many fine photos have been taken of rainbows passing in front of snow-capped mountains and into green valleys and red-rock canyons. But your own neighborhood has many a wondrous sight to see through rainbow light.

71. *Rainbow seen from behind.* One time I stared dozens of miles across the flat prairie and saw the setting sun break out and shine through rain that was falling from an approaching storm. The storm arrived at my location several hours later—too late for a rainbow. But the town I had come from was off in that direction, and I knew that people there must have seen a rainbow while I was staring at the sun shining through rain in the west (to them, I was beyond the rainbow). The next day, when I returned to the town, I found out that I was right. So, in a sense, I had been staring at the back side of a rainbow. Of course, I saw no beautiful colors. It might be accurate to say that a rainbow, no matter what angle you try to get on seeing it, has no back or side—only a front!

72. *A traveling rainbow warned-of beforehand.* You do not have to live on the Great Plains to experience something like the previous situation. In New Jersey I have called people to tell them that rain followed by abrupt clearing and a low sun were coming their way—and that, therefore, a rainbow I had just seen from a place already passed by the shower was also headed their way.

73. *Infrared rainbow.* Is there a rainbow in wavelengths of light other

than those of the visible spectrum? Robert Greenler proved that there is one in the near infrared by photographing it with special film and filter.

74. *Lysol bow.* The rainbow phenomenon can occur in substances other than water. I once accidentally discovered an unusual bow (with a fine secondary) in a spray of Lysol brand disinfectant in an otherwise dark room into which some sunbeams were shining. The major ingredient of Lysol is alcohol.

75. *Moon rainbow.* Although moonlight is immensely dimmer than sunlight, and the proper kinds of storms are less frequent at night, the moon is quite capable of producing a visible rainbow. Because of the lesser intensity of the moon rainbow, however, the eye can seldom perceive its colors. Do not confuse a halo around the moon with a moon rainbow (the latter, of course, occurs in the sky opposite from the moon).

76. *Blue-sky rainbow.* Is it possible to see a rainbow in a blue sky? Yes, the droplets can be slowly falling or floating as perhaps the last remnants of a cloud's dissipating.

77. *Eclipse of the sun corona rainbow.* On at least two occasions a rainbow has been reported caused by the pearly light of the sun's corona (outermost atmosphere) during a total eclipse of the sun. Such eclipses last for only a few minutes and occur at any given location on Earth an average of only once in several hundred years.

78. *Eclipse of the sun chromosphere rainbow.* The chromosphere of the sun is visible for only a few seconds near the beginning or end of a total solar eclipse. There is one reported case of the chromosphere's causing or plainly helping to cause a rainbow.

79. *Rainbow in front of the moon.* This would be visible only in the daytime when the moon, pale as china, could be seen through the colors of the three-times-wider primary bow. The section of bow and the moon would be moving at different angular rates and in (slightly) different directions, but the sight could endure for quite a while. I do not know anyone who has seen this enchanted sight, but a friend of mine beheld the next best thing.

80. *Rainbow with the moon resting against it.* Naturally my friend was without his camera at the time. But this must have looked like something out of fairyland that one could scarce believe photographable anyway!

81. *Rainbow seen (or photographed) through polarizing filter.* The light of the rainbow is polarized, and if you view or photograph it through a polarizing filter it can appear three times more intense relative to its background. Yet a friend of mine thinks that photographing a rainbow this way is unnatural, and cheating.

82. *Day-long and Hawaiian rainbow.* There are places in Hawaii where

it rains almost all the time. On a day near the start of winter the sun would be low enough to cause a rainbow all day long there. Hawaii is famous for rainbows. David Ludlum points out that even the University of Hawaii's athletic teams are named the Rainbows and that the Hawaiian word for rainbow is *anuenue*. He also quotes the following fine passage from Mark Twain's "Roughing It" about Hawaiian rainbows:

> Why did not Captain Cook have taste enough to call his great discovery the Rainbow Islands? These charming spectacles are common to all the Islands; they are visible every day, and frequently at night also—not the silvery bow we see once in an age in the States, by moonlight, but barred with all bright and beautiful colors, like the children of the sun and rain. What the sailors call "rain-dogs"—little patches of rainbow—are often drifting about the heavens in these latitudes, like stained cathedral windows.

I wonder if some of these rain-dogs were not actually "sun dogs"—that is, mock suns or parhelia—or parts of other halo phenomena.

83. *Volcano rainbow.* A very strange kind of rainbow also sometimes visible in Hawaii is caused by the radiance from volcanic eruptions at night. A discussion of a recent instance is described in the January 1986 issue of *Sky & Telescope,* page 89. Its observer, Richard Nelson, named it the "Pele Bow" in honor of the Hawaiian volcano goddess Pele. This bow is red, like the lava light.

84. *Nuclear-explosion rainbow.* The most horrible kind—a desecration of what a rainbow should be.

85. *Sub-atomic rainbow.* According to the scientist H. Moysés Nussenzveig, there has been found in sub-atomic physics analogies to the rainbow phenomenon. But none of us will ever see one.

86. *Geyser (or thermal spring) rainbow or mistbow.* I did not get a chance to look for this at Yellowstone National Park, but I did see a fine "glory" in steam rising from a thermal spring. I had thought that the droplets in the spray of these thermal features like geysers would be too small to form a colorful rainbow, but the contrary seems implied by this splendid passage from Edwin Way Teale's *Autumn Across America:*

> . . . in midafternoon, we came upon the Giant Geyser in full eruption, supporting twenty tons of water in the air at one time, sending its billowing clouds of steam and scalding spray 250 feet above the ground, while in the midst of this superheated fury shone the delicate, ephemeral beauty of a rainbow.

87. *Lightning rainbow.* Why should not bright lightning produce a rainbow? (There are some problems, admittedly—not the least of which is that

when lightning is very bright it is usually also close and high in the sky— and dangerous!)

88. *A rainbow vibrated by thunder.* J. W. Laine reported a rainbow whose colors' boundaries were lost, whose yellow grew brighter, whose space between violet and the first supernumerary disappeared—each time there was thunder. Exactly how such an effect could be produced has not been explained.

89. *A rainbow with someone you know in it.* You look across half a mile of flat plain or field. You see the house of someone you know in the rain and light producing a rainbow.

90. *Spider-web dewbow.* Robert Greenler writes of studying and enjoying these. The fact that the webs may be angled in many ways in relation to you and the sun introduces interesting possibilities.

91. *Jonathan Edwards' spurted rainbow.* America's greatest sermonist was New England's Jonathan Edwards (1703–1758), most famous for "Sinners in the Hands of An Angry God." He was a phenomenal child prodigy and penned a brilliant essay on both the science and philosophy of rainbows before he was thirteen years old. In it he mentions (along with much else) that one can fill one's mouth with water and "spurt" it out to produce a full rainbow!

92. *Jonathan Edwards' splash-with-stick rainbow.* Edwards also mentions splashing a stick in a puddle when the sun is low to produce a full, colorful rainbow. In actuality, both of Edwards' experiments must be exceedingly difficult to perform successfully (Robert Greenler mentions looking for rainbows in the drops thrown up by cars passing on a wet road.)

93. *Mock-sun rainbow.* These mock suns (see Chapter 3 for more on them) can be bright enough to cast shadows and probably to cause secondary mock suns. A mock sun rainbow would be at the same height as the primary rainbow caused by the sun but would be to either side of the customary rainbow, interlinking with it. I have never heard of one being seen, but parts of this necessarily dimmer bow may explain some of the unusual rainbows that have been reported.

94. *"The black rainbow."* According to Aden and Marjorie Meinel, this is a name (of what ill omen!) sometimes applied to moon rainbows because they are so dim.

95. *Steambow.* This is what a geyser bow can be, but consider the more mundane sources of steam in which it might be visible.

96. *Rainbow enlarged by optical aid in relation to earthly objects.* In a way this is only a trick, but it is a good one made possible by the rainbow's peculiar nature. All of us have seen those telephoto shots of a setting sun or rising moon swelled enormously in relation to an apparently tiny boat

or person seen silhouetted against it. The same trick can be played with
the rainbow, which behaves as though it were at infinity. You will see this
gimmick used to good effect in rainbow calendars each year.

97. *Eclipse of the sun "diamond ring" rainbow.* I have read no reports
that this has ever been seen, though it is no less likely to be sighted than
a chromosphere rainbow. The "diamond" is a first starlike piece of the
sun's blinding surface coming into view through a deep valley at the moon's
edge at the very start or end of a total solar eclipse. It can be very, very
much brighter than the full moon. How many people would be able to tear
themselves away from looking at the astonishing few seconds of the diamond
itself to look the opposite way and see the rainbow it was producing?

98. *Fireball meteor rainbow.* Many fireballs around the world each year
are brighter than the full moon. A fireball like this ought to occasionally
cause a rainbow that could move laterally while rising or setting, all in
counterpoint to the meteor's motion. Again: who would ever look away
from the object itself—a brief and stunning phenomenon—to notice its
fleeting rainbow in the opposite direction? Perhaps a person blocked from
direct view of the fireball.

99. *Multiple rainbows on planets of binary star systems.* There are probably innumerable planets with both rain (not necessarily of water) and two
or more suns in the sky. The possibilities of rainbows on such worlds may
be even richer than on our own!

100. *First rainbow.* A child's first rainbow. It brings to mind those very
famous lines of William Wordsworth:

> My heart leaps up when I behold
>   A rainbow in the sky:
> So was it when my life began,
> So is it now I am a man,
> So be it when I shall grow old,
>   Or let me die!

● ● ●

There they are: one hundred rainbows. I hope you will pardon me for
some that were not altogether legitimate. Remember there were other
entries into which I placed several separate rainbows. And there were some
bows I did not even mention: for instance, the rare white dewbow—apparently dewdrops are very seldom as small as the droplets in fog. And
since first writing this chapter I have learned still more kinds: the glass-

globe rainbow, the abnormally large raindrop bow, disjointed bows caused by both rain and seaspray—which makes a slightly smaller radiius bow— at the beach. And many more! Whatever you may think of some of the preposterously rare or odd bows included in this list, I believe you should have a better idea of what a rainbow is and also a greater appreciation for how to find an "ordinary" one and what to look for in it. (How foolish that word *ordinary* seems in connection with the rainbow!) I believe you should also now have some ideas about where to look for new expressions of the basic theme of the rainbow phenomenon.

Do not be put off by the rarity of many kinds of bows. If you think about them and where and when to look, you will see quite a few of them. You can bring substantially more primary rainbows, and other kinds too, into your life. To be even one rainbow richer is wealth inestimable.

In George MacDonald's fairy tale *The Golden Key,* a boy is told by his great-aunt that if he could reach the end of the rainbow he would find there a golden key. But, says the aunt, nobody knows what the key is for or will open—that he has to find out. The boy suggests that because the key is gold it might be worth a lot of money if he sold it. But the aunt replies, with simple and utter conviction: better never to find the golden key at the rainbow's end than to find it and sell it.

Every rainbow is a potent reminder of the beauty and mystery of the world. The rainbow exists for no practical purpose, and that is precisely its great value. The rainbow is pure beauty and wonder, and it reminds us— with a shock of awe and a chill of appreciation—that the highest purpose of all is to marvel and appreciate the endlessly deep richnesses, perfections, and mysteries of this world and what we can glimpse beyond it (and I do not here mean beyond in outer space). Whatever we do of "practical" nature can ultimately be important only insofar as it enhances our appreciation of and wonder for our lives. The rainbow is one of our great reminders—and proofs—of that truth.

5

A Night at Meteor Cove

WHEN I WAS in the early to middle years of grade school in the first half of 1960s, the Perseids introduced me to the pleasures and solitude, the patience and vast private wonder of meteor watching. To some people it might have seemed a strange and silly hobby to sit for hours in the dark staring at the sky in the hope of seeing "shooting stars." But such people did not know about *meteor showers*. They did not know what they were missing.

My secret hobby, a patient seeking and spectacular finding of literally moving beauty, was a "secret" all over the heavens of certain late hours of certain nights each year. I was not just hoping for meteors, I was seeing them: on the very best occasions, several hundred a night, in all colors, in brightness sometimes greater than any star's, in fragmenting and bursting and tumultuous flights, in zooming followed by seemingly phosphorescent trails lingering for breathless seconds after the disappearance of the meteors themselves!

The hobby or passionate avocation of meteor watching was as rare for adults as for children in those days, and it is perhaps even more rare now: the glare of our cities means that most people must travel many miles to get a good view of all the "falling stars" that a meteor shower has to offer. But the astronomical holidays of the major meteor showers—essentially the same nights each year—are clearly worth a drive. In the side yard of my childhood home in the country, I saw more than seventy, once even more than eighty Perseid meteors an hour in my best hours. Yet in later years there was one drive to see the shower from a special site that remains in my memory very vividly, even if it was only the night after the shower's climax, even though I therefore saw far fewer meteors than in some of the other, earlier, more private years. It was the night at Meteor Cove.

The night before that night was the one on which the Perseids reached their maximum. At this peak time, a careful observer with very clear and moonless skies many miles from city lights will usually see fifty, sixty, or more Perseid meteors in his or her best hour. That hour usually occurs well after midnight, when the spinning Earth is turning us right toward the meteors, and the spot they appear to come from—in the constellation Perseus—is high in the sky. But on the maximum night of August 11–12, 1980, the very hazy and mostly clouded skies where I lived gave little hope. I was watching a still partly unclouded area in the northern sky when I heard the first thunder.

Timing the lapse between flashes and thunders I found that the storm was already audible from at least fifteen miles away. The sight of the flash from lightning over a starry sky with a meteor flying through it is a strange, exciting one which I had beheld best one Perseid night several years earlier—after which a friend and I had driven down a lonely country road from our isolated site with eerie, sometimes towering "steam devils" coiling their funnels up in front of our headlights as they rose from the cooling but still hot-from-day road surface. But on this night in 1980 there were to be no meteors visible. At least not of the astronomical variety. The word *meteor,* you see, is from the Greek *meta* ("beyond") and *aeirō* ("raise") and originally meant any atmospheric phenomenon. In technical parlance, one still hears of various types of "igneous" or "fiery meteors," including lightning, and of various types of "aqueous meteors," including rain. There were certainly a lot of both of these examples that night. That night belonged to lightning. A friend on the open shore by the bay reported that lightning was bright enough that night to keep the photoelectrically controlled streetlights off for quite a long time with its false day. The final two and a half hours of the night, which should have been best for the Perseids, were a wild chain of one strong thunderstorm after another.

My only hope was that the clearing after the great storm passage might itself be vigorous, that the atmosphere would be rinsed and wrung clean for the best possible beholding of a still at least fairly good display of "shooting stars" the next night.

My hope was fulfilled, but not only with what I expected. There was to be more, and different.

•   •   •

What were these Perseids I waited for (and still wait for) each year, which I learned to hunt sight of while many other children learned to hunt (and cause death of) terrestrial prey?

Meteors are bits of rock or iron that enter our atmosphere from space at such great speeds that they burn up from friction with the air. It is usually only the light phenomenon (mostly incandescent air), however, that is called a meteor. The rock when in space is a *meteoroid* (smaller, merely by definition, than an asteroid); if found after having reached the Earth's surface, it is a *meteorite*. Very, very few of these objects do reach the ground. Most burn up completely fifty miles or higher above the surface. That is the fate of most meteors. And their origin? Some may be pieces of those worldlets— the asteroids or "minor planets"—that mostly rove the spaces between Jupiter and Mars. These are the ones with a real chance of becoming meteorites. But most meteors are originally the fragile pieces of matter that first appear in the dust tails of comets. Over thousands of years these *meteoroid streams* along the orbits of comets (or former orbits of comets) diffuse to a breadth of millions of miles—which Earth passes through each year on the nights we reach our intersections with the orbits, the nights of the meteor showers.

What is a meteor shower? It is not meteors (also called "shooting stars" and "falling stars") pouring from the sky as plentiful as raindrops (they are almost that plentiful only in the rarest instances—the awesome *meteor storms*).

Major Annual Meteor Showers

	Peak	*Above Max.*	*Some Visible*	*Hourly Rate*[a]	*Time*[b]
Quadrantids	Jan. 4	Jan. 4	Jan. 1–6	40[c]	6 A.M.
Lyrids	Apr. 22	Apr. 21–23	Apr. 18–25	15	4 A.M.
Eta Aquarids	May 5	May 1–10	Apr. 21–May 12	10	4 A.M.
Delta Aquarids	July 29	July 19–Aug. 8	July 15–Aug. 29	35	2 A.M.
Perseids	Aug. 12	Aug. 9–14	July 23–Aug. 20	50	4 A.M.
Orionids	Oct. 21	Oct. 20–25	Oct. 2–Nov. 7	25	4 A.M.
Taurids	Nov. 3	Oct. 20–Nov. 30	Sept. 15–Dec. 15	10	Midnight
Leonids	Nov. 18	Nov. 16–20	Nov. 14–20	5[d]	5 A.M.
Geminids	Dec. 14	Dec. 12–15	Dec. 4–16	50	2 A.M.

[a]Assumes clear, moonless sky far from city lights.
[b]Time when radiant is highest and sky still fully dark.
[c]Far more in some years for any given part of the world.
[d]Far more in some years, stupendously more at thrice-a-century intervals (next time is 1998 or 1999).

But a meteor shower is an increase in the typical six to ten meteors per hour that you might manage to see in clear country skies without moon on an average night. The meteors in a shower also are special in seeming to come from just one small region of the heavens (you might see a Perseid meteor appear in another part of the sky, but if you drew an imaginary line back from its path the line would go to a point in the constellation Perseus). This point or small area that the shower meteors seem to come from is called the *radiant*. If showers are caused by our passing through a stream of meteoroids traveling virtually parallel to one another through space, then why should these meteors appear to diverge from a radiant? The effect is merely one of perspective. It is similar to the seeming divergence of snowflakes from a spot ahead of you when you are driving your car through a snowstorm; or, more prosaically, the way that the parallel rails of a railroad appear to diverge from a single spot in the distance.

All of this explains what meteors are (pieces of rock from beyond our world, billions of years old, born of comets or asteroids, space visitors from millions of miles away!) and why they glow (at temperatures of at least several thousand degrees Fahrenheit) when they speed across our sky (at velocities of up to 150,000 miles per hour and more!). But what makes the Perseid meteor shower special requires further explanations.

I had long since (in childhood) learned those explanations as I woke (late) the next day to a clearing sky—and hope of a fine aftermath of the Perseid peak that night. The weather forecast held almost no chance of any cloud in the clear air of the coming night of August 12–13, 1980. So it was time to call some friends and to pack our lawn chairs for a trip to the house at Dyers Cove.

Renting that house for two available months was our artist friend, Pat. She had no phone there (purposely), but she had earlier told us that we were welcome to visit for the Perseids. It was already past 11 P.M. as we finally picked up the last person and headed away from town. Because the highest concentration of Perseids in space had been encountered by Earth the previous night or day, and the rates of Perseids usually fall off more quickly after maximum then they increase before maximum, perhaps we should have been out earlier—even before Earth had rolled us farther under Perseus and past midnight where we would be encountering more head-on, and therefore swifter and more numerous, meteors (more numerous with all other factors being equal, which they were not). But as Earthly lights fell behind us and the woods at last thinned and ceased, the opening night deepened over the marshes and around us—as our car deep-ended into it. We felt as though we had left behind that little garage that is all that most of man's work is and were driving, silently, out into the universe.

When we reached our destination, we could not get our doors open before all of us had seen meteors. I watched in amazement and admiration as a gentle speck of a meteor glided exquisitely through the tiny gap between the Great Andromeda Galaxy (itself splendidly visible to the naked eye) and the star nearest it.

In that secluded spot, there was a fresh, cool, gentle breeze off the dark waves, and as we walked on the beach with Pat the summer Milky Way arched over everything, completely from horizon to horizon—almost. For at the very southern end, below the bright "teapot" pattern of Sagittarius the Archer, below the distant star-cloud that lies in the direction of the hidden center of our galaxy, there was the awesome now-and-again glow of lightning within the cloud masses of a thunderstorm at the southern horizon. But it was not a danger. We sat there in the slight refreshing breeze getting more than a bit cool with the heavens perfect over all the rest of their grand extent—the storm must have been a hundred or more miles away, in another weather regime of the country, from which we were protected. It had been irrevocably driven from us by the cold, clear barometric hills of air upon which we now were lifted up under the stars.

Part of what explains the popularity of the Perseid shower is its season. For observers at north temperate latitudes where the northerly radiant in Perseus is well placed, this Perseid month of August is one of vacation weeks and mild nights—also significantly longer nights than those of June and July. Of course, thunderstorms are still a frequent hazard. And, in the entire eastern United States, haze can still at this time of year be as bad as in the earlier summer. But this time of year makes for much more endurable—indeed, often comfortable—hours of night sitting than those of the Perseids' strongest rivals for the title of "best shower," the Geminids of mid-December and the Quadrantids of early January. That period during which the Geminids and Quadrantids occur is also the cloudiest and stormiest of the year for much of North America.

The real essence of why the Perseids may deserve the title of best, however, is their combination of high peak meteor rate, long period of significant numbers, and relatively high dependability. The peak numbers of Perseids may be no better than of Geminids and the Perseids no more dependable (even a bit less, though they are also more likely to have a 100-per-hour maximum than the Geminids, perhaps). But the Perseid shower lasts longer. A few Perseids will usually be seen well before the end of July. And the best measure of the interesting part of these showers' duration is the time they are above quarter-strength (that is, at least a quarter the maximum number per hour is visible). For the Perseids, this period is about 4.6 days, for the Geminids only about 2.6. For the Quadrantids it is only 14 hours (an extremely narrow meteoroid stream in space). Only about once in 8 years does an observer at a given location have his or her longitude turned closely enough toward what used to be the constellation Quadrans Muralis during the few hours of Quadrantid peak activity—at which time 100 or more "Quads" an hour might be glimpsed.

The Perseids were recorded in China as early as A.D. 36 and in Europe at least as early as 811. There were sometimes spectacular Perseid displays recorded in these earlier times, and perhaps the shower's strength has waned a bit, or the numbers evened out over a longer period of hours and nights. Or perhaps not. The Skalnate Pleso Observatory reports that low Perseid rates in 1911 and 1912 stirred speculation that the shower was vanishing, but 1921 rates were tremendously high, and those of 1951 unusually good. Always there are individuals with perfect skies and skilled eye in a given year who run up impressive totals. I recall, I believe, 140 Perseids in an hour by one Harry N. Bearman one year in the early 1970s. But then in 1980—and for a few years after—reports of 100 Perseids per hour during a short peak were common, and rates of more than 200 per hour were neared or exceeded. Most experts concluded that the expected approach of the Perseids' parent comet, due between 1980 and 1984, was causing the dramatic increase.

They may have been wrong, but the topic of the Perseids' parent comet is another fascinating one. In 1866, Giovanni Schiaparelli (the famous observer of Mars) made the first link between a meteor shower and a comet— the Perseids and a comet that had appeared in 1862, Swift-Tuttle. This Comet Swift-Tuttle had reached second magnitude with a 25° long tail, but if its orbit is calculated accurately it has the potential to pass close enough to Earth to become one of the few most spectacular comets in history. Its true brightness may be somewhat greater than that of Halley's Comet, and therefore the brightest of any "short-period comet" (those that return within periods of less than 200 years). If it reached its closest point in space to the sun about August 12, it would get so close to Earth that it would probably outshine Venus and display a tail spanning the sky for several nights around that time.

Knowing all this, many of us waited for Swift-Tuttle with great anticipation, especially when the Perseid rates for several years showed such a dramatic increase. The comet was expected to return sometime between 1980 and 1984—and right there a problem was evident: such uncertainty in the orbital period.

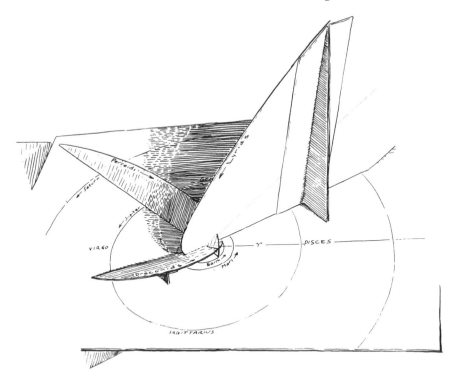

Three-dimensional diagram showing orbits of the Perseid, Lyrid, and Draconid meteor stream. Derived, with permission, from Guy Ottewell's *Astronomical Companion.*

Swift-Tuttle has not returned as of mid-1988, and Perseid rates have dropped back to about their typical levels. Although astronomers are confident in their associations of certain comets with certain meteors, this first identification of them all—Schiaparelli's link between the Perseids and Comet Swift-Tuttle—may not be correct. If Swift-Tuttle's orbit is even a little different from what was thought, it may have passed close enough to Jupiter (for instance) for that orbit to be changed and for us to never see it again. But if its orbit was not quite what we thought, then our linking of it with the Perseids and their orbit is incorrect, and another comet may be the Perseid parent. But the years ahead will be interesting ones as we see how the Perseids behave and whether Swift-Tuttle ever puts in an appearance—maybe in late 1992 as expert orbit calculator Brian Marsden thinks.

There on the beach at little Dyers Cove, each of us was seeing up to thirty to forty Perseids an hour at our best, and we believe that fifty-six different Perseids were seen by the combined efforts of all six of us in the first hour. Meteors other than the Perseids added as many as twenty more

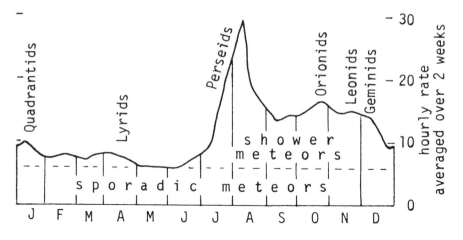

Hourly meteor rates during the year. Derived, with permission, from Guy Ottewell's *Astronomical Companion*.

to the hourly totals of each of us. Some of these were "sporadic meteors," belonging to no established meteor shower (thus coming from anywhere in the sky)—though perhaps most sporadics are really the last lonely survivors of a meteoroid stream (and resultant shower) that had been mighty in prehistoric times when its members had appeared in great numbers and its comet had blazed brightly. Now it had diffused into the common cloud of meteoroidal particles that we can see in part as the *zodiacal light* (Omar Khayyam's "false dawn"), the pyramidal glow in the east before true morning twilight or west after true evening twilight. Most of the non-Perseids that night at the cove, however, were not sporadics. They were Delta Aquarids.

The Delta Aquarid shower usually peaks in late July or early August when the Perseids are just getting started. It can produce rates of twenty, thirty, or more per hour. The strongest substream of the Delta Aquarids is above quarter-strength for about seven days, and this stream is plentiful enough for a whole month to make it altogether the most abundant shower (greatest total number of meteors during the year). Its peak, though, is more like a fairly high plateau, being less lofty and dramatic than the Perseids'. Delta Aquarids are very difficult to distinguish from Capricornid meteors, occurring as they do at around the same time and from nearly the same place in the sky; but they are very easy to distinguish from the Perseids, whose far faster streaks from the north contrast with the only moderate-speed Delta Aquarids (and Capricornids) from the south. It is a kind of shooting-star shootout (north vs. south) that produces a midsummer meteor maximum—never at any time of the year are meteors so numerous (this is another reason why Perseid nights are often favored over those of the Geminids and other showers that occur at different times of the year). This time of midsummer means (in the northern United States, at least) that corn is high and many berries are ripe for the picking—a time of culmination and fulfillment mirrored in the heavens' time of highest meteor rates. Shooting stars are then ripe for the picking, too.

From our reclining chairs we watched not so much the Perseid radiant as the regions some distance from there—the best all-around area for meteors. Farther from the radiant, meteors are generally fewer, though they do travel longer paths. Near the radiant itself the paths of meteors are very short—in fact, many simply flash out as bright points that are stationary if they occur very close to the radiant . . . a meteor coming right at you. One year I saw several Perseids near the radiant that looked like long, oblong balls of light, not moving, in which numerous brighter sparkle-points glittered for a second before the whole phenomenon vanished.

One Perseid we saw there at Meteor Cove left a bright yellow *train* (trail) whose glow was visible to the naked eye for a spectacular fifteen seconds after the meteor itself had disappeared (Guy Ottewell mentions a Perseid with train observed for a full minute and a half). These seemingly phosphorescent trails are actually the glow of gases ionized by the meteor's passage through the upper atmosphere. Trains are common with Perseids, as are *bolides* (exploding meteors) and *fireballs* (meteors brighter than Venus, the brightest point of light in the heavens). We did not see many bolides nor any fireballs that night but many tints and an indescribably rich array of meteor behaviors. Several Perseids traveled nearly parallel paths within a few seconds of each other. We experienced many lazy lulls without meteors and many dramatic spates of many meteors. The Perseids have a reputation for sometimes being irregular in this way. I do not think it is psychological. I once went ten minutes without seeing a single Perseid and not long after saw eight in another ten minutes.

As we watched in both peace and excitement, the hours slowly passed, the Earth slowly rolling us under the stars. How well that starfield of Perseus and Cassiopeia and Andromeda and their environs is etched in my mind—from that uninterrupted night at Meteor Cove, from my childhood years of Perseid showers. You will have to decide for yourself how often you want to watch meteors with other people, how often by yourself. But even in a large group of people, the meteors and especially the stars outnumber humans, and you may find yourself extended to the solitude of the heavens and filled with star-touched thoughts.

Wish on a meteor and you supposedly get your wish—I always find myself wishing for another beautiful meteor (and usually getting it)! But there are many legends and beliefs associated with meteors. The Perseids themselves have been called St. Laurence's Tears because his martyrdom was on August 10, A.D. 258 (how grimly appropriate to think of the Perseids as burning tears, considering that Laurence's death came upon a grid-iron.) Guy Ottewell has quoted Manilius: *Nunquam futilibus excanduit ignibus aether*—"The ether never blazed with futile fires"—and Guy quite rightly added, "That is to say: all meteors mean something." I agree. Every meteor means beauty, surprise, wonder, and a stirring of thought for its fortunate observer.

For most of that exquisite night, the thunderstorm at the south end of things, below the heart of the Milky Way, flickered a mystical accompaniment to the meteors. It seemed the cove was filling up with them at times— as the coves of our hearts were with their beauty. We were not only at the meeting place of sea, land, and cool air, we were where we could survey hundreds of miles—from meteor-filled space to stormland, several states away. We were where we could look across the universe not only of light-years but also of wonder—and even survey, in its light and peace, ourselves.

For more than five hours, with almost no break (certainly no going inside) we lay, enthralled by the performance of more than 300 meteors of many kinds. We had a chance to ride—in lawn chairs!—out on the edge of the world to witness nature in her full-force beauty and reality. We watched until, as always on the nights of the Perseids, the Pleiades had floated on high and Taurus and Orion—missed since spring—had followed. As always, Sirius, brightest of stars, had not risen quite early enough to be seen in the morning twilight—which this year spread pale at last over all the stars and meteors until flaming into the vivid pink of a Mt. St. Helens– enhanced dawn.

Two final and perhaps most important things I remember about that night:

First, I remember our discovery of the phosphorescent creatures on the beach! I still do not know what the minute creatures were, and I still long to go back and find them again. All I can tell you is that they shone like a miniature sky full of stars in a handful of dirt and sparked like a tiny Perseid storm of meteors about each footstep we took on the sand.

Each year that I go out to see the Perseids I hope to catch not only a glimpse of meteors but also of those nights twenty and more years ago when the child that was me sat out under the stars and watched the Perseids with freshest joy and satisfaction. That is the way every Perseid maximum should be—whether it is your first or your fiftieth. And, often as I look for the

Perseids, I remember the galaxies of phosphorescent creatures on the beach—plus something else.

Two months after the night of Meteor Cove, mere weeks after our artist friend had left there, one of the worst storms in years flooded the bay shore. We traveled down there afterward and saw that the little house at Dyers Cove had been carried a half mile or more inland and was broken into shambles. It was as if the storm in the south that meteor-filled night, held in abeyance for a while, had returned at last to destroy utterly. But if the fleeting fire of meteors can be so beautiful, their activity and boisterous bright death a lovely part of a night of moving peace laid as if in perfect crystal in the universe, then why not many other things' fire and death? Why not people's? The night at Meteor Cove is always there.

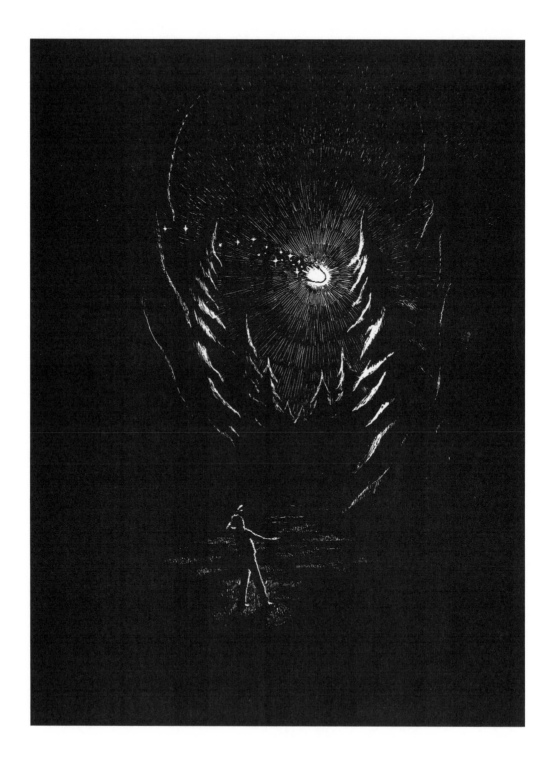

6

Flight

IT HAD SOMEWHERE BETWEEN about 4 years and 4 billion years of living—if roaming space unseen and almost eventlessly could be called that. It had exactly 10.5 seconds of dying—if making a half million hearts beat with fear and wonder in green light could be called that. It seemed to send out sound at the speed of light in a way that may have helped unlock one of nature's most puzzling mysteries. It remained visible for longer to more people than perhaps any object of its kind had for years. And many of those people, especially those from eastern Pennsylvania across New Jersey to New York City, are still wondering, years later, just what it was they saw.

I know. It was a green and red, hypersonic, probably electrophonic and meteorite-producing fireball meteor. As bright as a half moon.

I also know, with a fair degree of accuracy, after many conversations with dozens of eyewitnesses, after many months of my own figuring, after the five years later computations of an expert: how high, how bright, how fast, how loud, how fragmenting it was! And, very roughly, I know the most fascinating answers of all—I know where it came from and I know both where it ended and with what result or remainder (could some of its "pieces of the sky" be found?). I had no choice about whether or not to try learning these things. For, you see, I was one of the fireball meteor's witnesses. My memory of it would not let me rest.

This, then, is the story of one great fireball's flight of death and beauty through our atmosphere as I learned it. It is also, in a sense, the story of all great meteors and how they both frighten and ravish with beauty us their observers into thinking and learning about our universe. The story begins on a quiet summer night.

•   •   •

It was the evening of Tuesday, August 24, 1982. Had it been a Sunday, the number of people leaving the Jersey shore around twilight would have been phenomenally higher. Even so, the fireball could not have come at a much better time of year or any better time of day to be widely seen in this region.

The night was a calm and mild one there on the edge of the New Jersey Pine Barrens where I live. The last vestiges of twilight lingered dimly in the west, but the rest of the sky, especially the east, was already aglint with stars. Cirrus clouds were climbing and thickening in the southwest but so far had only just begun to reach the night's fine gathering of moon, Jupiter, and Mars in that part of the sky. The moon was just over a day short of first quarter, so its nearly half-lit form cast considerable light across open spaces in contrast to the deepening shadows of the trees in the yard and the already almost black walls of the surrounding forest.

The setting was beautiful and a little eerie but not really uncommon. I might have stepped out only briefly for a sample of the moon and planets that night if I had not been looking for Comet Austin.

I owe my view of the most exciting fireball I have ever seen to the modest beauty and interest of a comet merely at the edge of naked-eye visibility. Another night, I believe I glimpsed this comet without optical aid, but because it got quite low in the northwest before evening twilight was over those nights, binoculars were always necessary for any really proper view. In telescopes Comet Austin displayed several degrees of dim tail and a richly structured head. It was the brightest and most interesting comet of 1982, and many years have not had one as good. Little did I know that Austin would soon become for me an auxiliary object, almost a footnote to that night and to that summer of another object—this meteor visible only from a corner of one country not (like Austin) the whole world yet from my side yard visible for a few seconds stupendously.

I had not been out long and was just beginning my search for Austin with 7 × 50 binoculars. A problem was the high treeline to the northwest and the trees in the yard past which I had to maneuver my view. I am very fortunate that I was at that moment positioned near enough to a break in the trees that revealed a fairly large patch of the northeast sky.

I am also lucky that when the moment came my eyes were not clamped to the binoculars pointing near the legs of the Great Bear where Comet Austin glowed. Yet I wonder if the light that now appeared would have pried my attention out even around the edges of the binoculars' eyecups. I was standing with the binoculars hanging on my chest, gazing under a branch toward the comet's direction, when something—and of course it must have been the radiance—made me glance up to the northeast.

I had been a devoted watcher of the heavens for about twenty years— since rather early childhood—and I had seen many bright meteors before. I had seen quite a few fireballs—meteors even brighter than Venus—and even several fireballs on one night a few times. What now met my stupefied gaze, however, was rivaled in brightness probably by only one fireball I had ever seen, one almost as bright as a half moon I glimpsed rush down the heavens in an instant in the great 1966 display of the Leonid shower. But my view of that earlier prodigy had not been good, mostly because of its great swiftness and brevity. What now confronted me almost sixteen years later was so magnificently slow and enduring it was not a fleeting startlement, however great: it was an object with an awesome presence staring back at me and countless thousands of other people—staring with intense colors, with writhing tongues of flame, with steadily departing trail of "sparks" or "stars." The forest permitted me to see only a fifth or less of its flight's full duration, yet even that was several full, eye-cramming seconds and many more rapid heartbeats long.

The head of what I saw looked like a mass of green fire, rounded in front but widening backward to a fan of pointed tongues of flame. It was these tongues that were fiercely varying, and writhing more than a bit. The mass of green was at least several times less wide than the nearly half moon, but it was roughly as bright and therefore far more intense. The sky for several degrees around the head was whitish with strong radiance, and a whole section of the heavens was at least dimly affected by the illumination. If I had looked down behind me I would have seen my shadow cast in the green light, as so many other people did.

But no less beautiful and awesome than the head of the fireball were the pieces flowing from and following it. These looked like bright stars and were intensely red. The brightest was a fireball itself, for though seeming about the equal of Jupiter, it obviously must have been brighter to have looked so bright in the main body's glare. Something like four or five other pieces at a time were visible and appeared to rival some of the brightest stars—but again must truly have been brighter. All the pieces were flowing steadily but slowly out and slowly falling behind the main fireball. But they were also pursuing the beginnings of graceful arcs forward and slightly downward toward the horizon while the parent body, the green head, declined little from its *apparent* height in the sky. (As we will see, its loss of true altitude was actually quite considerable during the period I viewed it.)

These "stars" are presumably what some observers call "sparks" when they more rapidly fly off and fade out. When slower and more enduring, as in the August 24 fireball, they suggest pointlike gleaming bits of metal flying (but floatingly) from a hot object struck on an anvil rather than actual electric sparks. These fragments of the meteor were dimming out, fading from red to gray to invisible like embers before my eyes, but a few were lasting at least most of these two full seconds during which the majestic spectacle crossed the open patch of my sky.

To complete my description—of two seconds!—I must describe the mighty fireball's motion. I can say that it was nothing like the usual streak or zip of a meteor, whose apparent motion is much faster than even that of a speedy airplane flying quite low. It was no swifter than a plane whose movement we would call moderately fast (don't forget we are talking *apparent* speed here—planes we see are always much closer than meteors, and thus the *true* speed of meteors is incomparably greater).

But my problem is this: how can I properly convey the effect produced on observers by seeing such an object travel with that kind of motion? The best word to describe the motion is *majestic,* but a majestic mountain does not move and a majestic ship is not fiercely varying, nor can it be startling and brilliant in anything like the way the fireball was.

Add to this unique combination of image and motion what can only be called the fireball's overwhelming silence. Even as slow—relative to most meteors—as this fireball was, its brilliance and its impetuous fragmenting and flames naturally compelled the mind to expect accompanying sound. For the eye to be forced to its most open, active, and fullest while the expectant ear is gaping wide but utterly empty is a strange thing. Thus the silence of the fireball was profoundly troubling—and beautiful, because for silence itself to speak more eloquently than sound is a wondrous situation sometimes claimed in books but very seldom experienced so literally in life.

Of course, the sound from an object many miles distant (as glowing meteors inevitably are) should reach the observer—if ever—long after the meteor is seen (usually a minute or two later). But no one has much success in telling his or her instincts that. You never have a chance to, anyway, during the shocking seconds of a fireball observation. Hearing not a sound from a ferociously active and bright meteor is one of the oddest things you can experience (especially when the object is as enduring as the August 24 fireball). One of the few things weirder is hearing a sound, eerie and bodyless and somehow qualitatively different from any you have ever experienced, in the very moment when you are seeing a fire-ball rush by. That is a seeming impossibility (sound cannot travel at the speed of light!), yet there is incontrovertible evidence that it has been experienced—at a number of fireballs and, I was soon to find out, at this one.

Two seconds it crossed my patch of sky with its moon-bright but more intense green flame, its graceful trailing embers, its unnerving silence. Then it had passed behind the dense masses of the towering trees in the forest edge there near me—flame still burning, embers still flying, silence still held. I did not know where it had first become visible or where it was last seen—neither where in my sky (if there had been no trees in my way) nor where geographically (over what points in New Jersey or elsewhere in the United States). I did not know whether its final appearance to observers with a clear view was a simple vanishing or a spectacular shattering, whether any pieces of it could have reached the ground. I did not know—but I had to find out.

In the weeks and months that followed, most of my information came as a result of my weekly column on astronomy in the Atlantic City newspaper. The first phone call I got about the meteor was unsolicited, from a woman twenty miles away who knew about my column. It came about ten minutes after the fireball had passed. On the other hand, I got some very important information from the father of a boy at a school where I talked as long as eight months after the fireball. Not even now, by the way, do I consider the case closed. I think about the valuable report of *a 1911 fireball* I got from its eyewitness as an offshoot of the 1982 fireball investigation! But even though southern New Jersey turned out to be the best place from which to have seen the August 24, 1982, fireball, several sightings that were just about invaluable to my quest came from much farther afield—and arrived to me via SEAN.

It was an otherwise fruitless contact with a local planetarium (after calling several Coast Guard and flight service stations) that led me to the Smithsonian Institute's Scientific Event Alert Network—SEAN. A small staff of geologists collects information on each month's earthquakes, active volcanoes, and—with the occasional help of meteor experts—on fireballs and meteorite falls. The network consists of both observers (often amateur) and professional researchers around the world, and the staff's task is not just to gather the network's reports but also to combine, coordinate, and often elucidate them. The result is published in the monthly SEAN *Bulletin* (you can subscribe to it for about $18 a year, or report a very bright fireball, by writing to SEAN, National Museum of Natural History, Mail Stop 129, Washington, DC 20560). A few days after the fireball, I was talking to SEAN staff members, and they were already able to give me several exciting observations from places other than South Jersey.

A few of the observations were from experienced meteor observers who knew much about the true nature of these objects and were not deluded in their estimation of certain aspects they saw. The average person cannot conceive that a meteor's true brightness could outshine a whole city's, even a whole country's, or that its speed is typically as fast to even a jet as a bullet is to a jogger! Somehow the unexamined assumption in mind is always that the object's true brightness is comparable to that of a bright fireworks rocket, or maybe moderately brighter: so the fireball must be not much higher or father than a fireworks rocket, a few thousand feet away at most. The aspect of most fireballs that clinches the (incorrect) conclusion of such extreme closeness is the object's apparent speed: how could anything whizz across the sky in a second if it were not almost close enough to clip off roofs?

But this conclusion is wrong. *When you observe a point or mass of light in the sky, there is no way to tell its distance unless you already know its true brightness.* The object in question could be a fairly dim object very close or a stupendously brilliant one very far away: the firefly 60 feet from you looks as bright as a car headlight 25 blocks away or a fairly bright meteor 80 miles away or a planet 800 million miles away or a bright star 100 billion billion miles away. Yet the most common remark from someone who saw a bright fireball is something like: "I saw it go down behind those trees [or buildings]—it must have come down on the other side of my neighbor's yard!" The more cautious person will say it was the other side of *town.* He does not understand that a glowing meteor is essentially never closer than many miles away and even that only if it is nearly overhead. If you see a luminous meteor low in the sky (down near the horizon), it is even farther— perhaps 100 or 200 miles distant. People are disappointed (some absolutely refuse to believe) that the fireball did not crash into those woods behind the house. Otherwise, the reality of how high, fast, big, and bright a meteor actually turns out to be is far more impressive than the common conception of a slightly enhanced fireworks rocket (albeit one from outer space) passing near.

Needless to say, people should not be belittled for lacking this esoteric knowledge about the true nature of meteors. Meteors are, after all, originally denizens of the truly alien world of outer space. Meteors and meteorites are the single age-old, flaming, and solid bridge between space and Earth, between the unimaginable emptiness in which all things move and the fullness of our rich, accustomed world of roses and rivers and birds and breeze—and "shooting stars."

•   •   •

At least a few of the observations of the August 24 fireball were made free of the common misconceptions about meteors. They were given in terms of *azimuth* and *angular altitude*. Azimuth is horizontal measure in

The Lost City, Oklahoma fireball of January 3, 1970, photographed by an automatic camera station. Courtesy of Smithsonian Astrophysical Observatory. Dimmer lines are star trails. Note one-per-second breaks in fireball trail caused by rotating shutter.

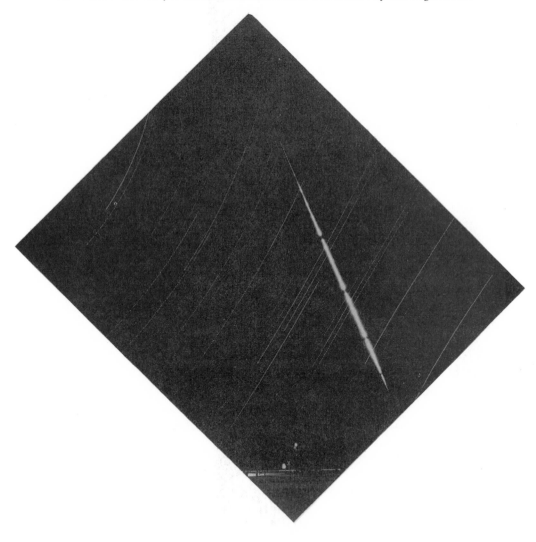

the sky with north as 0°, east as 90°, south 180°, west 270°, and so on back to 360° (a full circle around the sky), which is 0° and north again. Angular altitude must not be confused with true altitude, which can be given in miles, feet, or kilometers and which an individual observer cannot alone figure out for a meteor he or she has just seen. Angular altitude is vertical measure of the angle above the horizon at which something is seen in the sky. It is measured in degrees from 0° at the horizon up to 90° at the zenith (overhead point).

How easy is it to obtain an accurate azimuth and angular altitude for an object you have just seen in the sky? Getting a precise azimuth generally requires a compass; getting a good angular altitude requires a special pro-tractor or device like a "sky crossbow" (see the May 1981 issue of *Sky & Telescope* magazine for an article on how to make and use this convenient device). If you wish to obtain a rough angular altitude, you can take ad-vantage of the fact that a person's own fist held out at arm's length is about 10° wide (although large fists tend to be on long arms and small on short, you should try "calibrating" your fist by comparing its width—at arm's length—with the 10° distance between the two stars that form the top of the Big Dipper's bowl).

There are lurking problems for would-be obtainers of altitude and azi-muth. You may have to call your local airport to learn what corrections may be necessary to convert your compass readings to true, extremely accurate directions in your area (if you wish only to report your compass reading and not trouble yourself further you can of course do so, merely making sure to specify that your compass reading is uncorrected when you write to SEAN or contact any other authority collecting meteor sightings). A special problem for people trying to judge angular altitude is an instinctive tendency we all have to overestimate greatly the altitude of any object appearing relatively high in the sky. For instance, my initial impression was that the August 24 fireball was between 60° and 70° high in my sky; only later did I discover that it was a bit less than 50°, little more than halfway up the sky. To most people, anything halfway up the heavens seems virtually overhead!

Of course, these difficulties for meteor reporters are usually secondary to more fundamental ones that stem from the meteor's startlingness and brevity. Can you keep your wits well enough to fix where you first and last saw this sudden marvel in relation to prominent landmarks? You must be a little lucky even to have landmarks. If you are in a wide-open area you may get a good view, but there will be no towering tree or building near you to help gauge the angular height of the fireball. If you do have a large object near to use as a landmark, make sure you mark where you were standing when you observed the fireball (in such a case, a slight error in remembering where you stood could create a very large error in your altitude and azimuth estimates). Of course, ideally you might be able to fix a fireball's course in the sky very well by reference to the background of constellations it rushes through—assuming you know your constellations. Unfortunately, a very bright fireball usually drowns out all the lesser lights in the sky around it, often making such a plotting quite difficult. If the fireball appears in daylight, bright twilight, bright moonlight, or mostly cloudy skies there is little or no chance of using the background stars at all.

The best you can do is consider these problems now or outside tonight where you can test what you would do in just a few seconds of an imaginary fireball flight. Naturally, all such preparations will be far in the back of your mind when a bright fireball does appear. A person who takes inquiring walks under the stars or who looks at them not just admiringly but often, especially a person who looks at some of the year's meteor showers, is fairly likely to see at least one Venus-bright meteor that year. That one marginal fireball and the dozens (or hundreds) of less bright but often very beautiful other meteors are enough to keep a person fulfilled for years (especially when they are accompanied by thousands of sights of stars and planets and moon that meteor-interest gets one out more often to behold). And it is just a matter of living through a number of such star-filled and meteor-crossed years of nights to find yourself outside when a moon-bright fireball comes by. The vast majority of people, especially in our indoor, head-down age, will never see such a fireball in all their lives. But the person who enjoys being out and looking at the night fairly often is likely to see at least several such objects in the course of a life.

Yet it is almost inconceivable that a person should be expecting that mighty fireball on the night it comes. Most extremely bright fireballs are not seen in the meteor showers, which occur on the same dates each year—they are asteroidal in origin, whereas the showers are comet debris (almost always the very small dust grains originally released in comet's tails). An extremely bright meteor may be a quite small piece of matter coming in at especially high speed and steep angle (thus forced into ever denser atmosphere as quickly as possible and burning brightly but briefly). Or it may be one of the larger noncometary pieces, perhaps entering at lesser speed and/or at shallow angle. This latter kind of meteoroid, even when slow, has the potential to be so much larger that it has the potential to be a lot brighter. It also is capable of putting on a more complex performance and glowing longer—and is thus more likely to be spotted, remembered, plotted, and reported than a very bright but brief shower fireball.

What is the ultimate reason that fixing a fireball's path in the sky is so difficult? One moment you are casually scanning the heavens, or concentrating on some celestial object, or maybe not even looking to the sky. The next moment it is there. That is the reason. Your head spins and eyes bulge to fill with sight of a phenomenon whose true like you have never seen before and whose alterations may be as complex as they are strange and swift. It is one of nature's few most startling phenomena, and it will occur for only a few seconds just a few times—unexpected times—in your life. How will you respond?

In describing my experience of the August 24 fireball, I did not mention my thoughts and feelings in those moments, save for the eeriness and awe I felt at its combination of wild activity and brightness with relative slowness and palpable silence. I think, if I remember correctly, that I also had the discrete idea of grabbing the binoculars on my chest and raising them to my eyes for an enhanced view of the marvel. It is a good thing I did not try that—two seconds was not enough. But the other thought I know I had, even after years of watching meteors and thinking about them, was, in the back of my mind, that perhaps . . . perhaps this was some kind of fiery rocket launched from just a mile or two away and passing a dangerous few thousand feet from me! I am afraid the thought actually did pass through my mind. I quickly shook it off, but it may have had influence on the one action I recall doing after I turned to face the spectacle. Whether it really was the thought or just my body's completely instinctual reaction, I am not sure. But I do know that my hand flew up for a moment as if to ward off the danger of flaming and dropping embers from above that hole in the trees through which I stared up at the incredible fire.

Perhaps you can now understand why even the best observations of the fireball that I got from SEAN and from my own quizzing of South Jersey witnesses were not accurate enough to permit an *easy* solution to my most important questions: where had the fireball hurtled—how high and over exactly what territory and, most thrillingly, to what end? With several accurate enough sightings it is possible to use triangulation with protractor and straightedge on an ordinary road map to get a rather precise plot of the path. Corrections to the true altitude should be applied for the two facts that Earth is round and that gravity bends even the fastest meteor's course slightly more downward. But along the latter, lower part of a flight path (where observers usually best see it and we may get meteorites) these factors are usually small enough to be ignored (initially, at least). In practice, the problem of imprecision in the altitude and azimuth given by even some of the best observers is usually much more important. That was the case with the August 24 fireball.

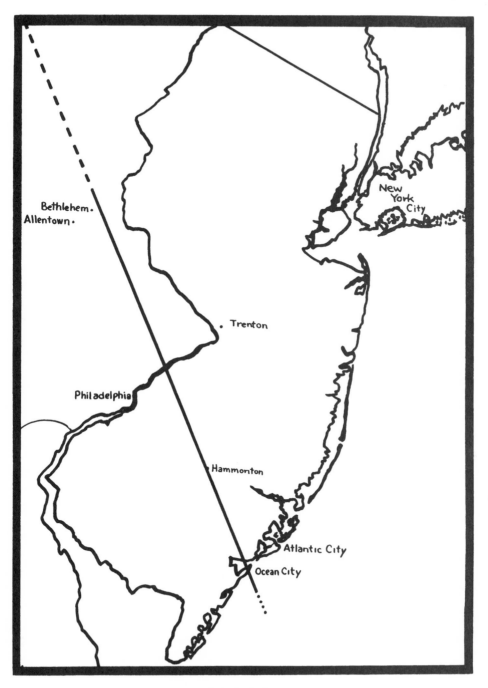

Flight path of the August 24, 1982, fireball.

I continued to struggle, juggling what were the best fits I could make by assuming various amounts of error by various observers. But certainly some marvelous facts became immediately apparent from these other observations. I did get a good ground track. And my two seconds of wonders from one place had been only a small sampling of all there had been to see and experience.

The most unusually fortunate circumstance was the fact that Fred Janicke, an experienced meteor observer, was looking west from Boonton (in northeastern New Jersey) at precisely the moment the fireball appeared—and was able to click the stopwatch he held in his hand at both the start and—10.5 seconds later—the end of the flight. In this time he watched the mighty meteor cross at an average rate of about 7° per second (it would have taken almost full four seconds to travel a length equal to that of the Big Dipper!) and end up all the way in his south where, having faded a bit (from its moving away from him), it had at last disappeared with a terminal brightening in Scorpius the Scorpion. Janicke, one of the devoted meteor observers of the Sheep Hill Astronomical Association, had seen the fireball appear not far from the bright star Arcturus.

Trees had blocked me from seeing how the fireball disappeared, but what had looked like merely a terminal brightening to Janicke from the opposite end of the state had appeared as an impressive final fragmenting to most (but strangely not all) observers in South Jersey and southeastern Pennsylvania. Those final pieces had faded from view in mid-air—so there was no chance of any of them reaching the surface? Not true! Contrary to popular belief, meteorites essentially never reach the ground luminous (unless they are the rare once-in-10,000-years giants). If a meteor is still fairly bright when it reaches an altitude of about five to fifteen miles, its disappearance means only that the denser air has slowed it down enough so that it ceases to glow. Soon after, the still denser atmosphere down lower will take away the rest of its original velocity, leaving it with only the speed of free fall that a parachute jumper would have: meteorites fall their last few miles as dark, cooling objects that finally strike at "only" a few hundred miles per hour. The observations I had gathered suggested that the fireball had been very low indeed—perhaps less than ten miles up—when it faded from view, with many fragments still as bright as Venus to one pair of observers at least several dozen miles away. The good news: this meteor, I thought, probably did produce meteorites. The bad news: observations from shore cities, especially Ocean City, New Jersey, suggested that the pieces from the final shattering must have reached the surface at least a small number of miles out over the ocean. The only hope for recoverable meteorites would be from "stars" that broke off a little before the final fragmentation—those earlier pieces might, I believed, have come down over land in South Jersey.

Most unusual of all in the reports I got from observers were the mentions of "anomalous sound." I must save a full discussion of this topic for another time. But I will say that a number of people, dozens of miles apart and dozens of miles from the fireball, heard sound simultaneous with its passage—a seeming impossibility (sound is tremendously slower than light) but one confirmed by expert observers at a certain (small) number of other fireballs. The best theory may be that a large meteor's passage disturbs Earth's local magnetic field, creating very low-frequency radio waves that might cause the distinctive "anomalous sound"—hissing or swishing—by directly stimulating the nervous system or brain in a way that mimics impulses from the auditory nerves. (A partly analogous case: rubbing closed eyes seems to produce the sensation of colors.) One of my observers, Wayne Smith, thinks that the sound he "heard" changed pitch—to the best of my knowledge a unique observation that may help solve the amazing mystery!

How many fireballs in the pages of the 1982 SEAN *Bulletin*s were in some way or other as grand as my mid-Atlantic coast fireball of August 24? Quite a few! Of course, to rate against one another these meteors, each unique in its awesomeness and beauty, is foolish. I did not lose my respect for the grandeur and rarity of the August 24 fireball by reading about others that were grand; instead, my respect for the richness of nature and the number of prodigies it can bring forth was greatly increased. I will say, though, that the period from July 1982 through January 1983 was—to my taste—better than any since in producing a variety of strongly colored, booming and bursting, wildly impetuous, at times even truly terrifying fireballs in the pages of the SEAN *Bulletin*. And I will also suggest that as you sample the following selection from these months consider that none of them had several of the August 24 meteor's most interesting characteristics or a quite similar combination.

July 12, 1982, southwest Michigan. This fireball is one of the few in years of SEAN *Bulletin*s that may have been visible even longer than the August 24 meteor. It burned for ten to fifteen seconds and went virtually all the way across the sky for viewers in Michigan. It was blue and red or red with "continuous sparking," and some witnesses heard sonic booms from it.

July 18, west Australia. An orange-red fireball already brighter than any star appeared and moved rapidly northwest. It disappeared; then, suddenly bright as a crescent moon and blue, it reappeared, still moving northwest but now faster. It was emitting fragments that rivaled the brightest stars and that changed from blue to red as they vanished. Finally the fireball exploded in a "brilliant violet-blue flash" almost as bright as the full moon, leaving three or four small fragments as bright as bright stars that continued on somewhat farther before disappearing.

A blue train (trail of ionized gas) about 10° long (twenty times the moon's apparent diameter) was visible to the naked eye for eight minutes (eighteen in binoculars) and during that time drifted on upper-atmosphere winds. It made "an S-shape after two minutes, two roundish clouds in four minutes, and a single cloud in six."

August 12, Norway. A purple and green fireball became as bright as a half moon in the half-second it lasted. It left a purple and green train that was at first brighter than any star; the train was visible for about one and a half minutes.

September 5, eastern Massachusetts. An observer at Martha's Vineyard saw a fireball move partway down the southern sky from near the zenith in about three and a half seconds. It was white and green with a thin green tail about 10–15° long, and it was as bright as the full moon.

September 29, southeastern Nebraska. Seen well before sunset, it was so bright it was dazzling even in the daytime sky. It was white, with a long tail leaving sparkling pieces, and it disppeared behind a cloud.

November 30. Just a few weeks after a meteorite crashed into a house in Wethersfield, Connecticut (see next chapter), a fireball brighter than the full moon zoomed low over the state. It was yellowish-white with a blue trail that persisted for about thirty seconds after the meteor burned out. "No sounds or meteorite falls reported."

January 3, 1983, mid-Atlantic states and midwestern United States. At 6:31 P.M., a red or orange fireball brighter than the full moon, traveling about north to south, was seen from eastern Maryland to central Indiana. In the Baltimore–Washington area, it was seen to pass below the horizon and *then* produce two red flashes visible after it had passed from view! In Blacksburg, Virginia, some observers saw it coming almost straight at them and therefore saw it appear stationary until it fragmented into ten pieces as if it were a "mid-air plane explosion." In West Virginia, people in many counties heard sonic booms and rumblings "about three to five minutes after the fireball lit up the sky." At the Greenbriar Valley Airport, the sound was likened to that of "a furnace blowing up."

This fireball may well have been visible from my home in South Jersey. I was looking in the right direction that evening when I stared off to see an amazing orange Venus setting in the second purple-light glow of a fine volcanic twilight—but I went inside about thirty minutes too soon to see the distant fireball.

• • •

What an incredible variety and virtuosity of colorful and beautiful, of roof-shaking and mind-ringing heavenly violence. Of course, all of this is merely a good selection from a certain six months—and you may have noted my list suspiciously dominated by U.S. meteors (because the interesting SEAN pages are—how many stupendous fireballs from other parts of the world in this period were never heard about by the Washington, D.C.–based SEAN staff?). If you want fireballs still more spectacular you have only to cast a bit farther over the years. The most extreme displays of fireballs truly stagger belief—even to hear of them, let alone to experience them. There have been meteors as bright as the sun—and probably brighter. Nor does the phenomenon have to be catastrophic like the Tunguska event of 1908 (quite possibly caused by comet, not meteor) to attain such awesome brilliance. At any rate, that is what some meteor reports suggest, if we can believe them. For instance, on November 4, 1977, parts of Ontario were witness to a fireball that shook buildings for fifty kilometers around and was said to be as bright as magnitude -26—virtually the sun's equal!

Another kind of fireball performance is practically as mind-boggling and eerie if sometimes slightly less riveting. I refer to the fireballs of extremely long duration. Those visible longest of all are big ones that happen to enter our atmosphere at such a shallow angle that they eventually escape back into space!

The most famous (but apparently not most recent) instance of an escapee occurred August 10, 1972, when a brilliant daytime fireball was seen, photographed, and even captured on movie film as it remained glowing over a ground path of about 900 miles from central Utah to central Alberta. Actually, the best plotting of its path and the certainty that it escaped were obtained by a U.S. Air Force satellite's infrared radiometer which detected the fireball's heat. The instrument found that this fireball got no closer than 36 miles from Earth's surface, over a point near the middle of its observed trajectory, in Montana. That result agreed well with the fact that sonic booms were heard only in Montana. This fireball was much less bright than the sun, even to the best-placed observers, but certainly more like the sun than the full moon in brightness to those it passed right over. It also produced green and blue flashes and orange sparks visible even in broad daylight, appeared to be spinning, and left a trail like a wide jet exhaust that remained visible for around 30 minutes. How long was the active meteor itself visible? Remembering that even a few seconds of a bright fireball may seem like minutes when magnified by awe, consider that some observers actually did watch this mighty object for more than half a minute. Though not very fast for a meteor, it must have traversed the United States from Mexico to Canada (though not visible until reaching Utah) in little more than 1½ minutes. A far more dramatic point is this: it was probably luminous for a total of 101 seconds! The only facts perhaps as remarkable as this and the fireball's escape are those relating to its possible size and weight. There have been many estimates of these; the largest (assuming the meteor was stony and not iron) is a diameter almost the same as the length of a football field and a mass of as much as 1 million tons! An impact from this object would have produced an explosion of atomic bomb intensity. (By the way, if the object that exploded over the Tunguska region of Siberia in 1908 was a meteor, it may have been *twice* as wide as the 1972 escapee and may have weighed 7 million tons—about the weight of the Great Pyramid of Egypt, but moving at probably more than 20,000 miles per hour.)

Could a fireball itself (not its trail) remain visible from one location for much longer than a minute? Not a single meteor. But on February 9, 1913, the Great Meteor Procession was numerous groups containing many bright meteors each and traveling one after another in slow majesty across the skies from Saskatchewan, Canada, to the South Atlantic—where they were seen still burning as they passed on out of view for shipboard observers. This means the Procession was visible for almost one-quarter of the way around the Earth, during which time, it has been estimated, the meteors lowered only from about thirty-five miles high down to thirty miles high. John O'Keefe in 1961 suggested that these meteors were fragments of a small asteroid that had been temporarily circling Earth as a second moon. Like artificial satellites with which we are now familiar, the body's orbit may have at last decayed enough for the Earth's atmosphere to capture it in a fiery but very shallow-angled and long-lasting descent. O'Keefe even thought the Procession may have survived for several fiery revolutions around the turning Earth, but he failed to find any sightings along the predicted track the Procession would have taken after its first traversing of the United States took it from Buffalo, New York, to Sandy Hook, New Jersey (unfortunately, much of the heavily populated area of this first and only proven track of the Procession over the United States was cloudy that remarkable evening). There remain problems for all the interpretations of the Procession's behavior so far advanced, and that is not surprising considering the confused observations of so complex and astonishing a spectacle. But at the very least, we know that the Procession remained visible from individual locations for well over three minutes (it was about one hundred miles long) and was luminous along its track for at least the better part of an hour.

In all of these accounts, I have not yet described any fireball that reached the ground and was recovered as a meteorite—arguably the most exciting thing even a fireball could do. The topic is too rich. It deserves its own essay and gets it—the next in this book.

But now it is time at last to reveal what I learned about the size, speed, altitude, path, origin, and destiny of the great New Jersey fireball. Much of what I learned I figured out for myself roughly, and was later confirmed by the computer program of Dr. David Meisel, president of the American Meteor Society. But it was not until the very last month that this book was being written, almost five years after the fireball, that Meisel and I conferred closely enough for him to determine several important corrections of my estimates (for instance, of the object's likely mass)—plus determination of one thing else I had no inkling about: the fireball's original orbit as a meteoroid in space. That orbit proved so unusual and interesting that it even made the other facts about the fireball seem almost customary by comparison! I will therefore save the revelation of it for last and tell the meteor and meteorite part of the tale first.

At a few minutes before 8:33 P.M. EDT on August 24, 1982, a meteoroid made of rock or iron is about to encounter the Earth. It swoops in over the daytime shoulder Earth is gently turning away from the sun, trying to catch up with both the planet's orbital speed and rotational spin that is carrying the east coast of the United States deeply into twilight. The meteoroid is roughly the size of a large television set and weighs very approximately 400 pounds—hardly enormous in the popular mind, though you would be hard pressed to budge it if you found it in your back yard!

But even though it is slower than many meteors relative to Earth, the probably somewhat irregularly shaped body is traveling at a velocity of approximately 49,000 miles per hour (perhaps as much as 63,000) relative to the planet's surface—dozens of times faster than jets, bullets, or sound. If it were to hit even a large building at this mass and speed, the destruction of the edifice—perhaps of a whole city block—would be complete.

At about 8:32 P.M. EDT—one minute before it becomes luminous—it is about 400 miles high over southern Canada, possibly in some strong Northern Lights, though these and the meteoroid are still in sunlight at their great altitude above the twilit land below. If the meteoroid could see it would now have a view from the Arctic to Mexico, from the Atlantic to the Rocky Mountains. As it passes near the eastern end of the Great Lakes, all of those lakes except Superior have lost the sun (the sun is setting in Chicago).

The gases of the upper atmosphere are astonishingly thin, but when the meteoroid has reached an altitude of about ninety miles, heating of it from its vigorous impact with the gas molecules is beginning.

Finally, at an altitude between 72 and 54 miles, even an atmospheric pressure about 0.000001 of that at sea level has been enough resistance to send the object's temperature in a few seconds up to 500°–1,000°C.— and to turn it into a meteor. One moment nothing is visible even from the spot roughly 60 miles straight below the meteoroid, about 10 miles north of the Allentown–Bethlehem, Pennsylvania, area. The next moment an envelope of incandescent air is glowing green and so bright that it is seen from Virginia to New Hampshire, rivals the half moon in brilliance, casts radiance that throws sharp and moving shadows in the landscape of three states at a time.

The fireball races at a scarcely diminished velocity toward South Jersey, a luminous orange trail stretching for about five or six miles behind, which is about six to ten times the apparent diameter of the moon as seen by Fred Janicke four to five dozen miles away.

Just a few seconds later, the fireball has passed from the coldest to the hottest part of Earth's atmosphere and at about thirty miles altitude in the ozone layer is crossing over the Delaware River just north of Philadelphia, not many miles south of where George Washington did. It is now low enough to cause normal sound that might be heard at ground level a few minutes later (though I have no reports of this). But the "anomalous sound" *is* beginning (radio waves generated by the meteor-disrupted geomagnetic field?) to be experienced by a few observers thirty or forty miles ahead near the southern New Jersey shore, who now all see the fireball climbed to about halfway up their northwest or north–northwest sky. Major fragmentation could have begun several seconds earlier, but it is about now that the pieces are getting far enough from the main body to be seen (they themselves are probably the sizes of pebbles and fists), and the main body is (perhaps) near maximum brightness.

The main body itself finally begins to slow drastically and in a few more long seconds is around twenty miles over Hammonton, New Jersey, as I first glimpse it. Apparently its green envelope of incandescent air has grown immensely since Janicke saw it as a point object. Now it is about a third the apparent diameter of the moon for the closer observers—an umbrella of flame many city blocks wide soaring over the South Jersey pine forests at the rate of about half a county a second. Red fragments trail the main body by as much as two miles or more.

Another one to two seconds, and it is much lower than the highest point Chuck Yeager reached with a jet from the ground before the jet's malfunction and his famous fall of nearly twenty miles. The fireball passes behind trees from my viewpoint but a fraction of a second later is a little more than ten miles due west of Atlantic City. David and Michael Capizola in Newfield then are seeing a brightest of all—seemingly Venus-bright—fragment coming back from the maybe fifteen-mile-high main body.

One more second (its next to last) and the main fireball is thirteen or fourteen miles over Ocean City, New Jersey, still hissing with "anomalous sound." It has reached the shore of the Atlantic Ocean about the moment the Capizola piece, maybe one or two miles behind it and lower, fades from visibility.

Finally it is over the ocean, its speed down to somewhere between about 20,000 and 30,000 miles per hour, or 5½ to 8½ miles per second, but perhaps considerably slower. Only about 7 miles south–southeast of Ocean City (maybe less), the pressure on the body finally shatters it. Its altitude is then about 12½ (plus or minus 3) miles high. This is roughly the height of the very tallest thunderstorms, about right in the lowest of the El Chichón volcano layers of haze, about twice as high as Mt. Everest, well above airliner altitude but not that of many other jets. A viewer 75 miles away in Longwood Gardens, Pennsylvania, sees about 4 pieces in the final breakup; from Millville, New Jersey (34 miles away), it is about a dozen pieces; from Northfield (13 miles), again about a dozen pieces, perhaps because the person is looking more nearly straight down the flight path. Perhaps the best view I later learn about is had by two women in Heislerville (one of whom I meet a month and a half after the fireball, for other reasons, and who later becomes my wife!). They are about 26 miles away from the final breakup but are viewing it from an excellent side angle. They see about two dozen pieces, many of them as bright as or brighter than the great planet Venus: a shower of fireballs, a shower of Venuses!

And then the night is dark and still. The eerie "anomalous sound" ceases maybe at the moment of, or just after, the final shattering, just before the fading out of the downward arcing final pieces. The night is still—except for the gasps and exclamations of many thousands of people in a number of states (but especially South Jersey).

But what had happened to the fragments after they ceased to be luminous? Meisel's program indicated a considerably more likely than not probability of meteorites, but how much of the original mass of roughly 400 pounds would be left? Answer: probably no more than a few pounds! Of course, these bodies were lost to the ocean (though Meisel suggests it is not impossible that they later washed ashore!). But what about the "Capizola piece" and other earlier fragments? There is a significant chance that a few of them came to solid ground somewhere along a stretch of a few

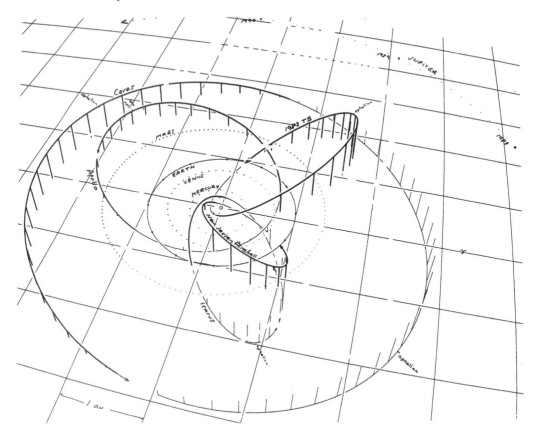

Three-dimensional diagram showing orbit of New Jersey fireball. Generated by Guy Ottewell.

dozen miles under the flight path. This is far too large an area to search—yet they are probably lying out there at this moment!

I still wonder if I might someday get a few more key observations to help find them —even so many years after the event. But as far as meteorite recovery goes, I think about a meteorite in South Jersey that I heard was discovered something like a hundred years after its nineteenth-century impact (estimated from laboratory analysis). That could well be the fate of the 1982 fireball's possible fragments. Perhaps someone in the twenty-first century will guess which meteor such a meteorite was derived from.

Now comes the *really* unusual part of the story! The origin of the 1982 fireball looked for a while (before we refined the analysis) as if it might have been artificial—not a meteoroid but a manmade spacecraft (or part of one). A small number of objects reentered that day, and the good people at Goddard Spaceflight Center are not told their time of reentry, only their day, by the American defense network NORAD. But fortunately the satellite possibility was soon eliminated by Meisel, and I also found out from Goddard that nothing in the proper mass range to be the fireball fell from Earth orbit that day. I write "fortunately" because for me it would have been unfortunate—interesting but aesthetically and scientifically less exciting— if the mysterious, glorious object had been reduced to some unpleasant fragment of "space junk." Instead what Meisel's program came up with (he feels quite reliably) was a natural meteoroid orbit—and one of the strangest ever determined.

I wrote earlier that the bright fireballs which penetrate very low are thought to be all asteroidal in origin. In the next chapter you will find that the three recovered meteorites which have had their orbits precisely determined all turn out to have been following courses quite similar to those of the Earth-crosser asteroids. Such asteroids and their little brother (but still sizable and very occasionally ground-reaching) meteoroids are believed to have been thrown out of their original orbits in the "asteroid belt" between Mars and Jupiter by one of several possible processes. They may eventually have found themselves in a belt orbit that brought them into a relation with Jupiter (and the other planets?) that kicked them into a new orbit across the paths of Mars, Earth, maybe even Venus. Another likely cause is collisions of asteroids in the belt. Such events must be exceedingly rare, but they must happen. Whatever the original cause, we may expect the orbits of Earth-crosser asteroids and meteorite-producing meteoroids to be generally very little inclined to the plane of Earth's orbits (and those of most of the planets) and to have aphelia (points farthest from sun) out in or near the asteroid belt with perihelia (points nearest to sun) rarely much farther in than Venus.

But Meisel finds that the Ocean City fireball (as we can call the 1982 object) had an orbital inclination between 34° and 44°, an aphelion between about 1.5 and 1.9 astronomical units (an "A.U." or astronomical unit is the average Earth-to-sun distance), and a perihelion between 0.1 and 0.25 A.U. (the range of these values reflecting a range of possible entry velocities). Those figures are remarkable! Canadian astronomers were able to calculate accurate orbits for many fireballs photographed from several locations by their camera network (see next chapter). In the full list of 50 such fireballs— all of their fireballs which were expected to have produced at least 50 grams of meteorite—the largest inclination is 24.6° (only 10 are over 15°), and the smallest perihelion distance is 0.405 A.U. (only 3 are less than 0.5 A.U.). The Ocean City object's perihelion distance was truly amazing—it was located far within the orbit of Mercury, much farther than any of the Canadian (or perhaps anyone else's) meteoroids and rivaled only by the asteroid Icarus and the strange asteroid or comet remnant (1983 TB on our diagram) Phaëthon, which seems to be the parent of the Geminid meteor shower—though both Icarus and Phaëthon have a much lower inclination than the Ocean City object.

The meteoroid's aphelion distance is perhaps more like that of Icarus than that of Phaëthon, but possibly closer than either to the distance of Mars from the sun. Could Mars have played a role in the Ocean City object's orbital history? The recent finding of a meteorite in Antarctica that could be a tiny piece of Mars raises the speculation that the Ocean City body (or its parent) could have been blasted off Mars by some past tremendous meteorite impact—though this, I hasten to add, is a very remote possibility indeed.

One thing seems certain: a body of this size could not have been in an orbit taking it so close to the sun for very long nor have lasted in such an orbit much longer. Either the meteoroid was swayed into that orbit (period or "year" somewhere between 0.8 and 1.2 Earth-years long) within recent times (a few centuries or even a few years ago), or it was until recently part of a larger body which endured in the orbit longer before the fearfully hot perihelion passes finally brought about a breakup. The meteoroid or its parent or grandparent was formed with the rest of the solar system about 4½ billion years ago, probably in the asteroid belt, and may have been swayed or knocked out of there millions of years or mere decades ago—but nothing less than a far, far larger object could have survived the perihelion passages in the final orbit for thousands of years.

It is difficult or impossible to choose certainly between the several exciting alternatives for the recent history of the meteoroid. But there is absolutely no doubt what would have happened to it in a matter of a few more orbits and years if it had not encountered Earth on August 24, 1982: even if the object was iron, its heating up to about 1,100° or 1,200°C. at each perihelion passage would have soon broken it up. When it hit Earth it was forty-six days away from its next perihelion passage (on about October 9, 1982), and that could well have been its last. It had to die, and die soon, whether or not it hit Earth, so we are lucky we got to see its death in the most spectacular fashion—and learn about its nature and life.

The dark and empty path where the fireball passed still seems to glow and reverberate in my thoughts—especially on summer nights similar to the one on which it occurred, most especially on every clear anniversary of that unforgettable evening. But if what we have determined about the orbit and mass of the Ocean City meteoroid is right, and it may have been part of a larger object, then my going out to scan the heavens in the early evening every August 24 (and the early evenings for several dates around that time) is not mere sentiment or a salute to memory. There is a distinct possibility that other pieces, *perhaps a few much larger ones,* exist in that same orbit and will encounter some part of Earth—perhaps your part—one of these years on or around that key date.

Even if no other piece so large ever encounters us, we might in some years experience a number of lesser meteors around August 24, all coming, as the great fireball did, from a radiant at about 8.84 hours of right ascension and +46° declination—right at the Lynx–Ursa Major border, near the Great Bear's foremost paw (formed by the close-together stars Iota and Kappa Ursae Majoris). In late August that point is very low in the north–northwest at nightfall for observers at 40° North latitude and then sets—but begins to creep up (a little higher than it was at nightfall) in the northeast just before morning twilight. Thus night's start and end would be the best times to look for a meteor shower that, because Iota and Kappa Ursae Majoris's proper names are Talitha Borealis and Talitha Australis, respectively, could be called the Talithids ("Talitha" is from the Arabic word for "third" because these two stars—northern (Latin "borealis") and southern (Latin "australis") were imagined to be the third set of footprints in the Three Leaps of the Gazelle).

If the Talithids are ever seen, they are probably not distant cousins of the famous Geminids—even though the Geminids have an "argument of perihelion" very similar to what the Talithids would. But if you or I see the Talithids some August night, what we will know certainly is that they are the siblings or children of that 1982 fireball whose 10.5 seconds of flame, fragmenting, "anomalous sound" and probable meteorite production have already led me on five years of adventure—and will never be forgotten as long as its thousands of observers live or this book is read.

Child and Willamette meteorite

7

Go to Innisfree

T.S. ELIOT WROTE in "Little Gidding" about the dove breaking the air with a fire of incandescent terror. That famous passage could be applied to the terror and wonder that is a great fireball meteor. But there is something a fireball can do that is potentially even more thrilling than flaming as bright as the sun or booming louder than a hundred thunders.

It can become a meteorite.

Meteorites are the secrets of the solar system's origins being dropped into our laps. But these pieces of the heavens, or pieces of the puzzle of the heavens, are even a good deal more than that. They are house-crashers, people-hitters, world-shrouders, evolution-alterers, Earth-pockers, rock-rainers, wilderness monsters, objects of ancient (and continuing) veneration, and the most exotic prey of all for some of the most patient, perceptive, and rare of Earth's people—the meteorite hunters. The story of meteorite falls, finds, and finders is a tale of highest romance fallen (quite literally) upon often humblest settings and people. As such, it is reminiscent of lines from another, probably even greater twentieth-century poet:

"I will arise and go now, and go to Innisfree . . ." But there is a far more specific connection between William Butler Yeats's ever-fresh poem "The Lake Isle of Innisfree" and meteorites.

The connection is that name: Innisfree. Only three meteorite falls have ever been recovered after meteor flights photographed well enough for the objects' original orbits to be precisely calculated. Like all falls, the three were named for places near them and, like many, just happen to have names that sound phonically appropriate, as well as mysterious and suggestive: Pribram . . . Lost City . . . Innisfree! Not Yeats's Irish isle, but the little town of Innisfree in Alberta, Canada.

Let Innisfree be a code name for all the romance of meteorites we can go to. Are the silver apples of the moon, the golden apples of the sun, and the rest of Yeats's "Song of Wandering Aengus" much more poetic or magical than the wanderings of the great meteorite pioneer Harvey Nininger through the plains, deserts, and (for meteoritic purposes only) even the bars of America? What tales, factual or fictional, are more gripping and full of exotic intrigues than that of the lost, vast Chinguetti meteorite with its finding in the Adrar desert dawn and the poisoning of its Bedouin betrayer? One of the eeriest and most adventurous true-life stories ever told must be that of Robert Peary's quest to bring back from the Greenland wastes the largest meteorite ever returned to civilization—the monstrous, sacred Ahnighito, which helped finance his historic first trip to the North Pole but almost cost him his life more than once. What kind of object caused the beautiful geometric crater most of a mile in width out in the Arizona desert? The seventy-ton shower of space rocks at Sikhote Alin? The eighteen-foot-deep hole on the outskirts of Kirin, China in 1976? The death of the dinosaurs? The bruises of Mrs. Ann Hodges of Sylacauga, Alabama, a person who is so far unique in all of history?

These matters are all episodes in the high romance of meteorites, which is now our subject.

When in December 1807 more than 300 pounds of meteorites fell in Weston, Connecticut, President Thomas Jefferson was indignant about the report of two Yale professors on the fall. "I could more easily believe that two Yankee professors would lie," he reputedly said, "than that stones would fall from heaven." Yet in this assessment Jefferson, who was quite knowledgable about astronomy, was speaking in accord with the view of most scientists of his day. It had been only four years earlier that the French Academy had finally accepted the cosmic origin of meteorites on the grounds of overwhelming evidence in their member Jean Baptiste Biot's report on the great shower of meteorites that had fallen at L'Aigle.

Strangely enough, many ancient Greek and Roman writers had believed that there were really falls of stones from the heavens (the first steps of modern science had been backward in this matter). Of course, the ancients' opinion was nothing but a hunch, for, as D.W.R. McKinley drolly observed, they were quite "unhampered by scientific knowledge." Perhaps the most famous meteorite fall of ancient Greece was one the size of a cart that landed in Aegospotami in Thrace in 467 B.C. (Pliny later wrote that Anaxagoras had predicted the day of the fall and had claimed that the stone had come from the sun.) Gertrude and James Jobes mention a version of the tale of Hercules in which his foe the Nemean Lion is said to have fallen from the moon ("in the form of a meteor," the Jobeses write). Livy wrote of a fall of stones in Rome around 654 B.C.

A chariot-sized stone is supposed to have killed ten men in China in 616 B.C. Meteorite falls and finds in China date back to before 2100 B.C. (though this is before the historical period), and iron meteorites were alloyed with bronze to make weapons there before 1000 B.C. Formal records on meteorite falls and speculation on their cosmic origins began in China as early as 645 B.C. Three hundred and sixty-five Chinese meteorite falls and finds before A.D. 1911 have been catalogued, with thirty-three instances of serious or calamitous damage caused by falls.

Even earlier than the Chinese use of meteorite iron seems to be that of the Hittites, usually credited with the first successful smelting of iron, around 1400 B.C. A list of treasures of a Hittite king in the sixteenth century B.C. mentions "black iron of heaven from the sky." More than three milennia later, the fantasy writer J.R.R. Tolkien tells the story of a hero with a sword "made of iron that fell from heaven as a blazing star; it would cleave all earth-delved iron." This sword, called Anglachel, was later renamed Gurthang ("Iron of Death") and replied in the affirmative when the tragic hero Turin Turambar asked it if it would help him take his own life.

Meteorites have been worshipped by many cultures. The black stone of the Mohammedan shrine at Mecca may be a meteorite, and they have been found in both an Aztec temple and North American Indian burial mounds. The cause for building the Temple of Artemis at Ephesus, one of the Seven Wonders of the ancient world, may have been a meteorite. There was at least allegedly still a meteorite there when the apostle Paul came. In the Revised Standard Version of the Bible (Acts of the Apostles, 19:35), the town clerk of Ephesus asks: "Men of Ephesus, what man is there who does not know that the city of the Ephesians is temple keeper of the great Artemis, and of the sacred stone that fell from the sky?"

Even though the ancients generally believed that stones could fall from the heavens, it was not necessarily understood that these stones were associated with the streaks of light in the sky called meteors. In the previous chapter I discussed how virtually all meteorites hit the ground a full minute or two after the last light of the meteor is seen—a fact that makes the failure to note the meteor/meteorite connection more understandable. On the other hand, Pliny did indeed write that the meteor (that is, the envelope of incandescent air around the object) was larger than the meteorite in the 467 B.C. fall. But in modern times, the idea that meteors sometimes produced meteorites did not struggle to life until late in the seventeenth century. Not long after, in 1714, Edmund Halley published a paper called "Account of Several Extraordinary Meteors," referring to two bright fireballs that had each been observed from several different places, permitting Halley to use triangulation and show that the objects were so high that they must have come from outer space. But no one seems to have paid much attention to the paper. Like many of Halley's ideas and works, this one was far ahead of its time—apparently too far.

When the idea that meteors could become meteorites was restored by science, the popular conception of fiery impacts and sizzling meteorites grew. And along with that usually incorrect idea came another: the bigger and faster a meteoroid in space and the meteor in the air, the more likely it would be to reach the ground and become a meteorite.

At first glance, the idea seems to be valid. A meteoroid's energy of motion, its kinetic energy, would be what would get it through to the ground, and this is merely the product of its mass times its velocity. Earth's atmospheric gases have far less resistance than a metal but there is a tremendously greater amount and distance of them to pass through than the door of a bank vault. According to Fred Whipple, a meteoroid weighing just 0.4 gram (maybe too small to see in your hand) would, at a typical meteoroid speed, hit the hull of a spaceship with the energy of 100 grams of TNT, the amount in the average hand grenade. Yet that same meteoroid might burn in our atmosphere at no more than about first magnitude brightness—and have no chance whatsoever of becoming a meteorite. Thus a few yards of lead or steel stopping a bullet would seem rightly analogous to a few hundred miles of atmosphere stopping the interplanetary bullet called a meteoroid.

Would that things were that simple! In reality, the fate of a meteoroid entering Earth's atmosphere depends on the mass, velocity, angle of entry, composition, peculiarities of breakup (one fragment may shield another), and other factors, all of which are not just variable but also interdependent. The atmosphere is a much more complex and changeable medium to pass through than any steel shield, too. Would the less-fierce friction on a meteor coming in at a shallow angle and therefore being more gently slowed before reaching the thicker air, would this improve such an object's chances of reaching the surface? Maybe. The longer journey (in both miles and seconds) would also allow more time for it to burn up. Whether it will become a meteorite or not will depend not just on its mass, velocity, angle of entry, composition, shape, and so on: it will depend on the *combination* of all these things—which itself is radically changing during the meteor's brief flight. Almost any one property a meteoroid has may be beneficial or harmful to its chances depending on what the other properties are. About the only exceptions are the most extremely small and most extremely large mass (always good for producing a meteorite—or, in the former case, micrometeorite), and very large velocity (critically bad for any but the afore-mentioned two kinds of meteoroid).

A further illustration of the situation's complexity comes from looking at this term *velocity*. It really covers a number of things. There is the true velocity of meteoroids in space, which is actually not much of a variable (most meteoroids in the vicinity of Earth's orbit travel at roughly the same true velocity). What is important is rather the velocity *relative* to Earth. A head-on encounter produces a greater total velocity, of course. And the most important factor in determining the relative velocity is usually one thing: the time of day at which the meteoroid arrives.

The time of day at any spot on Earth or in its atmosphere depends, obviously, on what part of its daily rotation that spot is at. The hours after midnight and especially around 6 A.M. are the time when a place on the spinning Earth is turned to face as nearly forward as possible into the direction of Earth's orbital motion. Meteors occurring at these times are consequently faster, hotter, brighter, briefer, and far less likely to survive than evening and especially 6 P.M. meteors. There are fewer meteors in the early evening because not all of the meteoroids catch up to the part of the planet being rotated away from them and many of those that do come in come in so much more slowly they do not burn bright enough to get noticed. If you want more numerous and brighter meteors, go out well after midnight; if you want longer-lasting and potentially meteorite-dropping meteors, go out well before midnight.

Of course, going out at even the best time of night for them is still not going to make the odds of your seeing a meteorite hit near you better than a tremendously long shot. According to one estimate, about 100 million to 200 million visible meteors occur in Earth's atmosphere *each day*—but only a few dozen produce meteorites. The smallness of many meteorites and the vastness of Earth's surface area—most of it sea, much of the land uninhabited—means that only a few score of these fresh meteorites are found each year, even if we count numerous individual members of the half-dozen or so falls that are located. The prospects are a lot better for you to discover old meteorites that fell in past centuries or millennia. A few hundred of the old ones may be found each year, or more if certain areas on the margins of Antarctica are searched for meteorites concentrated in those areas by the glacial flows. Roughly 3,000 meteorite falls (some containing numerous individual rocks) have been catalogued since such efforts began in the early nineteenth century.

These figures all suggest that there is little hope that you or I, even with some skill and luck, will ever find a freshly fallen meteorite. On the other hand, our chances of seeing a few meteorite-producing fireballs in our life

Barringer Crater in Arizona

are very good indeed. And if we have observed such a fireball's path carefully and alert authorities quickly, we may play a significant role in helping
other people locate a fresh meteorite—even if that meteorite is not in our
back yard but in another state.

So it is not such a bad idea to go out in the hours just after dusk with
at least the secondary (back-of-the-mind) purpose of trying to glimpse a
meteorite-producing fireball. The chances of seeing such a fireball in early
evening are extremely much better than in the last hours of the night. Time
of day is not the only factor, since Earth's tilt with respect to its orbital
plane and direction of motion makes your latitude and time of year significant also. These latter two factors determine how close you will get to the
precise apex or antapex around 6 A.M. or 6 P.M., these places being the
frontmost part on Earth in its orbital flight (apex you are near around 6
A.M.) and the rearmost (antapex you are near around 6 P.M.). How important are the various factors to the possibility of a meteorite? A study
by Ian Halliday and Arthur A. Griffin calculates that meteorite falls should
be a whopping eleven times more common at the antapex than at the apex
(the opposite—more at the apex—would be true for a planet with a thin
atmosphere). Outside the tropics, the variation is about 30 percent from

the beginning of spring (most meteorites) to the beginning of autumn (least meteorites) for each hemisphere. Latitude itself does not make as much difference as it would without the powerful effect of Earth's gravity in altering the orbits of meteoroids near it, but the poles should nevertheless have only about 85 percent as many meteorites as the equator.

Thus Earth's tilt, spin, and flight, along with the density profile of its atmosphere, all combine to determine how much on various parts of its beautiful body it receives the delicate touches of meteorite impacts.

Two other variables that play roles in a meteoroid's fate deserve special note: the composition and the type of vaporization.

Meteoroids that survive to become meteorites may be divided into three most fundamental compositional types, on the basis of their proportions of silicates and metals. There are the stony, the iron, and the stony-iron meteorites. The iron meteorites are much more likely to survive the passage through Earth's atmosphere (the fact that far more stony meteorites are found indicates that in space stony meteoroids must tremendously outnumber iron meteoroids). But there is a fourth type of meteoroid that may become a meteor but perhaps never a meteorite: the very fragile, partly silicate meteoroid derived from comets. None of these objects, including the members of all or most of our meteor showers, has ever been known to survive and become a meteorite. Presumably if there are meteoroids of cometary ice that enter Earth's atmosphere, even large chunks of comet nucleus ice, they too will not survive to reach ground. Meteorites are thought to be all derived from asteroids—with the possible rare exceptions of a few that may be pieces of the moon and the inner planets blasted off into space by impacts on these worlds.

Distinguishing between stony, iron, stony-iron, and cometary meteors is not always possible, yet clearly the fate of the object is affected greatly or crucially by this factor.

Besides making a distinction between meteoroids of different composition, we should also distinguish between several kinds of vaporization. Meteors suffer "ablation," the scoring off of parts of their surfaces by vaporization or by melting and vaporization. But small, short-lived meteors are consumed by vaporization directly from solid to gas, whereas large, longer-lived ones get a chance to have melting on their surfaces and so also have liquid-to-gas vaporization going on. The latter process actually helps cool the body below the surface, facilitating the passage to the ground even better than the simpler kind of ablation. (By the way, it is ablation that causes the beautiful depressions often called "thumbprints" on the outside of meteorites.) The question of how to calculate meteoroid vaporization is much further complicated by the existence of more or less of an envelope of incandescent air and vapor, its size and importance depending once again on most of the other variables of mass, velocity, angle of entry, and so on.

Certainly no simple table or graph can give us a meteoroid's chance of survival as a function of so many variables. But a classification of the fates and effects of meteoroids arranged in order of increasing mass (the most important and fairly straightforward variable) is possible and useful—if we are willing to accept certain simplifications and bear in mind that the results will be only very approximate. This classification is useful because it gives us a better grasp of the actual interplay of variables, or at least the results. It is also beautiful because, in addition to this grasping, the progression of fates—and effects on our senses and our Earth—is full of strange and marvelous surprises. A meteoroid can help a fifty-mile-high cloud to shine, shroud Earth in darkness and cold for months, shower the length of a small country with countless numbers of rocks, lie on the ground in a monstrous, awesome form to which the superstitious will pray, bury itself three-men-deep in a hole no wider than itself, blast open a crater many thousands of times larger than its own body yet symmetrical and beautiful in form. Let us examine how and when the play of various factors gives rise to these different possibilities. Two final qualifications: the following scheme is especially simplified by being limited to only very sturdy meteorites, and is increasingly speculative as we increase the mass of the meteoroid into ranges seldom if ever observed in history.

1. *Micrometeoroids in the range of a picogram (one-trillionth of a gram) to a microgram (one-millionth of a gram).*   Much too small to see individually even in your hand. They never become meteors but (presumably) always become meteorites—or, rather *micrometeorites*—once they enter our atmosphere. Their mass is so slight that even the exceedingly thin gases at high altitudes is sufficient to slow them before they heat up enough to glow or vaporize. They can drift for a long time in the upper atmosphere and probably form the condensation nuclei for vapor that freezes on them and that when concentrated in great enough numbers can be seen as the beautiful silver-blue and gold "noctilucent clouds." These clouds shine in deep twilight over high latitudes at a height of about fifty miles (the altitude of the coldest part of Earth's atmosphere). Micrometeoroids are eventually carried by air currents into the lower atmosphere and can be collected in rainfall.

The next four classes of meteoroids are described in terms of their meteors' brightness because they are the only ones you are likely to see as meteors more than once or twice (if that often) in your life.

2. *Produce telescopic meteors.*   According to David Meisel, meteors as faint as fifteenth magnitude have been seen with very short luminous trails. The mass and size of these objects are not much larger than those of micrometeoroids. All, of course, are vaporized completely.

3. *Produce faint naked-eye meteors.*   Grains up to a gram or so that would still scarcely be visible in your hand. They stand no chance of penetrating even much below about forty miles altitude before being totally vaporized.

4. *Produce bright naked-eye meteors up to the dimmest fireballs.*   Some of these bodies may be pebble-sized and weigh a number of ounces. Only if such an object is a fragment from a larger body breaking off in the lower atmosphere might it survive to become a meteorite.

5. *Produce fireballs ranging from slightly brighter than Venus to brighter than the full moon.*   Mass range from less than a pound up to a ton (potentially more, but limited to a ton for our purposes here). A small but significant percentage of these survive to become meteorites. Most meteorites come from objects in this initial mass range (which is usually reduced to a tiny percentage of the original mass) and therefore survive in this manner: they are decelerated until they cease glowing between about five and fifteen miles altitude, then further slowed until they lose all of their original "cosmic velocity"—only a speed of a few hundred miles per hour (because of Earth's gravity) is left to them when they thump or plunk to the ground as cool, dark bodies, burying themselves as much as several feet deep if they are large and the ground is relatively soft.

The following classes are of such rare occurrence that most people will never see even the brilliant fireball (brighter than the full moon) caused by one. We are fourtunate that no meteoroid of the final class has ever entered Earth's atmosphere in recorded history.

6. *Mass of 1 to 10 tons.*   These objects are so massive they generally retain a little of their cosmic velocity and are almost certain to survive. When they hit they may bury themselves deeper than smaller meteorites but still produce craters little wider than themselves. Only a few giant stony meteorites have withstood the pressures of entry at high velocity well enough to resist fragmentation and to be recovered as bodies in the 1- to 2-ton range. The old champion was the Norton, Kansas, meteorite of 1948 (2,360 pounds; hole 9 feet deep in a wheatfield). Its successor was the Kirin (or, according to a different spelling, Jiling) meteorite of China, which fell on March 8, 1976 while the astronomers of the world were watching bright Comet West. Kirin is a city of more than a million people, and this fall on its outskirts was therefore the most observed and (after interviews of thou-

sands of eyewitnesses) the best documented in history. The main fragment of the Kirin stone weighs 3,900 pounds and buried itself about 18 feet deep.

7. *Mass of 10 to 100 tons.*   These may always reach the ground (in part, at least) and do so retaining a significant fraction of their cosmic velocity. The result is that they hit the ground at 4 kilometers per second (roughly 9,000 miles per hour) or faster and cause an "explosion crater" wider than they in the act of vaporizing some of their own mass upon impact. Some scientists believe that no meteorite larger than about 100 tons will ever be found on Earth because although some certainly strike the ground while larger (even tremendously larger) than this, most of their mass will be vaporized upon impact by the conversion of some of their enormous kinetic energy into thermal energy. The largest meteorites so far found (always of the iron variety) weigh fractions of 100 tons and probably hit sandy or snowy terrain or landed along a slope similar to their angle of entry. Examples, like the Ahnighito meteorite, we will examine in a moment.

8. *Mass of 100 to 1,000 tons.*   May retain a majority of their cosmic velocity, but even if iron often break up in the atmosphere into smaller fragments. The result may be a major *meteorite shower,* like that of the object which showered the Sikhote Alin mountains a few hundred miles north of Vladivostok on February 12, 1947. The fireball was so bright that it produced prominent shadows in broad daylight and left a cloud, visible for hours, which is estimated to have contained about 200 tons of meteoritic material. The fall area itself was an awesome scene. There were 122 craters in the ground, 17 of them ranging from 10 to 26.5 meters (about 33 to 88 feet) in width. The deepest holes were about 20 feet deep. Numerous meteorites were embedded in trees, some even after richocheting off the rocky ground. About 23 tons of meteorites were collected, but as much as 70 tons were estimated to have fallen. The largest recovered fragment of this great iron meteorite weighed 1.9 tons.

9. *Mass of 1,000 to very roughly 10 million tons.*   These objects retain almost all or essentially all of their cosmic velocity and hit with so much kinetic energy that they are less likely to survive even as sizable pieces on the ground than objects of the previous two classes—in fact, they may even be less likely to reach the ground at all! The explanation lies in their great speed. Their kinetic energy may be so enormous that their "impact" with the atmosphere below about 10 miles causes some of them to vaporize as readily as objects of the previous two classes do when hitting the ground. Presumably the initial speed (therefore, time of day) and the angle of entry (if we ignore the key factor of the compositional type) would determine whether the virtually total vaporization would take place at the ground or

the lower atmosphere. Several probable or possible examples of the latter have been observed. There is even speculation that an especially large example of such an object was responsible for the famous and controversial "Tunguska event." Zdenek Sekanina, a comet expert, has rejected the idea that the fiery projectile that exploded so spectacularly over the Tunguska region of Siberia in 1908 was a piece of a comet's icy nucleus. Sekanina believes instead that the object was a meteor—a meteor whose diameter and mass have been estimated at more than 100 meters (longer than the length of an American football field) and about 1 million tons (equal to the weight of several of the world's largest oil tankers, fully loaded—but this object traveling at not a few tens of miles per hour, rather tens of thousands of miles per hour!). One calculation (now questioned) suggests 7 million tons. The Tunguska event occurred not long after 7 A.M. local time, so presumably the object came in at an especially high velocity. Perhaps the meteor that produced the 4,200-foot-wide, 570-foot-deep Barringer Crater in Arizona arrived in late afternoon or early evening, and thus was slow enough to survive until it struck the ground—and then probably vaporized almost entirely. The Barringer object was once estimated to have had a mass of as much as 15 million tons, but later calculations suggest that it may have been considerably smaller than the Tunguska body (if the latter was indeed a meteoroid). The velocity is the key. If it did arrive when Arizona was near Earth's antapex, then its velocity could have been something like 16 kilometers per second (roughly 35,000 miles per hour) when it hit and its diameter have been "only" about 80 feet, with a mass of a mere 65,000 tons. Yet T. R. LeMaire writes that Barringer Crater is an excavation of 300 million tons of rock that could hold two dozen Rose Bowl football stadiums and seat about 2½ million people. It was quite an explosion. How often do objects like those at Tunguska and Barringer occur? The Barringer Crater is often estimated as about 25,000 years old, though a few researchers have guessed that it is less than 10,000 years old. Almost certainly no fresher meteorite crater of its size exists anywhere on Earth. Perhaps Tunguska-like events are more common. Eugene Shoemaker calculates that a meteor with energy equivalent to Tunguska's occurs about once every 150 to 600 years. But perhaps other combinations of velocity, composition, and angle of entry may make it more likely for such objects (many perhaps faster but not so massive as that of Tunguska) to vaporize more gradually and higher in the atmosphere. Our uncertainties about objects of this size are great—but at least we may be comforted by the fact that there is no evidence of any other Tunguskas in history.

10. *Mass something like 10 million tons and up.* Such an object, measuring from ⅛th of a mile to 10 or 20 or more miles to diameter, would

deserve to be called an asteroid. Asteroids in this size range pass within a million miles of Earth maybe once or twice a century (the 2-mile-wide and more than 2-billion-ton asteroid Hermes passed only 480,000 miles from Earth in 1937). There would certainly be no possibility of the object's vaporizing before striking the surface of Earth. Its impact would cause a crater many miles wide and would directly endanger the lives of much or all of a continent before shrouding the whole planet in dust or water vapor, and therefore darkness and cold for weeks or months (the terrible "nuclear winter" that could be caused by a comet's entering our atmosphere, a giant meteorite impact, or a relatively small exchange of nuclear weapons). Earth's surface is three-quarters water, but an ocean impact might be worse, with mountain-high tsunami, water vapor shroud, and most extreme weather disruption resulting from what would be a generally more efficiently destructive use of the awesome energy. How often do meteoroids of this size and mass strike? Perhaps the best study in recent years figures that an asteroid with a mass of greater than 1 billion tons strikes Earth about once every 300,000 years. The dim remains of truly gigantic meteor craters, called "astroblemes," do indeed exist around the world. The cosmic origins of most have been disputed, but the meteorite explanation has been strongly argued for craters like the 25-mile-wide Vredefort Ring in South Africa and the very much larger Sudbury Basin in Ontario just north of Lake Huron. The latter is believed to have been formed sometime around 1.7 billion years ago. Still larger astroblemes have been suggested. One of the smaller, now completely underground at Redwing Creek, North Dakota, is a mere 5 miles across. Its date of origin, estimated as between 140 and 230 million years ago, raises a question that has been raging in recent years: what destroyed the dinosaurs and much other life on Earth about 60 million years ago? Was the cause astronomical? The Redwing Creek meteor fell in a time and region rich in dinosaurs, but they survived (admittedly the meteor may have been too small to have caused a severe enough Earth-shrouding). Proponents of an astronomical cause for cyclic mass extinctions of Earth species generally favor comets as the culprits. But why should they be more disastrous than the impacts of the largest meteoroids, which we believe to be very many times more frequent? Would the shroud created by a comet's dust and gases in our atmosphere be much worse even if we assume that no piece of the mostly icy comet nucleus ever survived vaporization to strike the surface? Our understanding about these matters is extremely limited— but still the fascinating debate continues.

The romance of meteorites has no stories more gripping and colorful than those of the giant meteorite finds. The very names of these monstrous

iron masses of strange form sound appropriately powerful and formidable: the Nullabor or Nullarbor (also called Mundrabilla) of desertest Australia, the Bacubirito of Mexico, the Hoba West and Mbosi and Chinguetti of Africa—and the Ahnighito of Greenland.

The adventure story of the last of these must be one of the most unusual and exciting in history. Its major characters were Robert Peary, who later gained historic fame for being (allegedly) the first man to reach the North Pole, and a family—a family of giant sacred wilderness meteorites. The Eskimos of the great Greenland ice waste thought the legendary meteorites they called the Man, the Woman, the Dog, and the Tent were members of a family group. And in a sense they may have been: they were possibly all huge fragments of one still more enormous body that had fallen from outer space.

In 1894, Peary finally succeeded in bribing one of the natives to take him to the Eskimos' mysterious source of iron. After a treacherous journey, they arrived and found it—the 2,500-pound meteorite called the Woman. The 1,100-pound Dog was not far away. But Peary could not learn where the Man was. A 20-ton meteorite found in Greenland in 1964 and now on display in Copenhagen's Mineralogical Museum has been called the Man, but according to one source, Peary's Eskimo guide claimed that there was a largest meteorite of all (even larger than the Tent meteorite) far inland.

But what of the Tent or Ahnighito meteorite?

In 1895 Peary floated Woman and Dog out to his ship on ice blocks and then the next year returned to capture the greatest prize, the Ahnighito. His ship was a steamer with icebreaker bow, he brought along hydraulic jacks capable of lifting 30 tons each, and the location of the Ahnighito was only 300 feet from the water surrounding its island in Melville Bay. But the Ahnighito was 10 feet tall and 7 feet wide. Peary was in for the battle of his life. Each time the jacks upended it, the iron behemoth would crash down, sending sparks flying from the rocks. Iron chain was squished by it. Buffeted by severe snowstorms, and fearing the bay would freeze hard enough to trap his ship all winter, Peary gave up the fight for that year.

Peary returned the next summer and succeeded in loading the meteorite—after it was almost dropped into the water when mighty waves from an ill-timed iceberg avalanche in the bay crashed into the ship. Early winter was again closing in, and Peary's ship almost did not succeed in smashing its way out. And still the monstrous meteorite that the Eskimos and Peary's superstitious sailors feared seemed to battle him: the immense iron mass played havoc with the ship's compass, and Peary went off course, struggling into a Canadian port burning his last ton of coal.

Finally, on October 2, 1897, Peary's ship reached Brooklyn Navy Yard.

To haul the meteorite down the streets of New York required eighty horses. Peary sold the Ahnighito to the American Museum of Natural History for $40,000—in those days a small fortune, and in this instance enough to help finance Peary's historic successful attempt to be the first to reach the North Pole.

It was not until 1956 that a special scale was brought in and the Ahnighito weighed. It turned out to weigh less than many people had thought—a mere 34 tons. That was still far more than the largest meteorite ever discovered in the United States, the 15½-ton Willamette. This iron meteorite had been found in Oregon in 1902 by a farmer who with ingenious engineering managed to haul the meteorite the 4,000 feet off of Oregon Iron and Steel's property onto his own—with one horse, one son, and one year of work! His scam was soon found out, though, and it was the company which sold the Willamette to the American Museum of Natural History, where it joined (and can still be seen with) the Ahnighito. No meteorite

Man and Ahnighito meteorite

looks stranger than does Willamette, with its enormous interior cavities on one side. But the Ahnighito remains the largest meteorite ever transported, the largest on public display, and second-largest ever certainly identified on Earth.

The only meteorite known to be larger is much larger and has never been moved from its impact site. The sixty- to seventy-ton Hoba West meteorite is located in Namibia (formerly called South-West Africa), where it mysteriously burrowed only about three feet into the native limestone.

There are, naturally, rumors of larger meteorites. The most notable is that of the Chinguetti meteorite, which only one European ever saw—if it was not a hoax. Captain M. Ripert claimed that he overheard some of his Bedouin soldiers talking about the meteorite and finally convinced one to take him to it. The guide consented only if Ripert brought no compass and they rode by night. According to Ripert, the two of them rode their camels 10 hours and arrived at the site around dawn to see the first light glittering off the sandblasted-silvery surface of a meteorite 330 feet long and 65 feet high! Ripert took with him a sample piece that later was proven to be a true meteorite, but his Bedouin guide died soon after they returned—supposedly by poisoning—and no one has been able to find the enormous object since that dawn in 1916. Was the whole affair a hoax perpetrated by Ripert? There are certainly some suspicious details to the story, and it seems unlikely that a football field–sized meteorite would remain hidden to this day even amid (or temporarily under) the dunes of the Adrar desert of Mauritania. There is also the theoretical consideration that so large a meteorite should have vaporized almost entirely upon impact. Still, could it be sand dunes with the proper slope that explain in part the fact that most of the world's largest meteorites have been found in the world's great deserts—this, and snow drifts in the cold desert of the Arctic? (Obviously the dryness of these regions has been a key to the meteorites' preservation for what is usually many thousands of years.) Perhaps a shadow of a doubt must remain about whether an object like the Chinguetti meteorite could · really exist.

No matter what the world's largest meteorite is, the fact remains that excitement and human drama surround the finding of all of the world's giants. The stories are too numerous to tell. But one thinks immediately of the many great meteorites of Mexico, whose chief is the roughly twenty-four- to thirty-ton, three feet by thirteen feet, and somewhat bent Bacubirito! And there is no more colorful story than that of the twelve-ton Morito meteorite, mentioned by the Spanish as early as 1600. Later in its history, a blacksmith attempted to cut it up for its financial value and heated it with a great fire. But the heat re-radiated by the mighty body was so tremendous that the blacksmith succeeded in chiseling off only a three-pound piece. Humbled, he surrendered and inscribed on the Morito the words that can still be read on it today at its resting place in Mexico City's School of Mines. The words capture well the awe that all of the world's great meteorites have rightfully inspired: *Sole Dios con su poder/ Este fierro destruira/ Porque en el mundo no habra/ Quien lo puede deshacer* ("Only God with his power/ Will this iron destroy,/ For in all the world there shall not be/ Anyone who can undo it.")

Few of us will ever be transported by night to a strange rendezvous in the Mauritanian desert or dare the crashing of the Arctic icebergs as winter begins in August. But we can all look for meteorites at home and in the places we do visit. The idea of a stone or iron from space crashing on the outskirts of Anytown, USA (maybe just off Main Street, maybe among Farmer Wilson's cows), has its own excitement, so much more unexpected than in the exotic wildernesses of Earth. And we may find that emulating the character of the humble and devoted lifelong meteorite hunters of our homeland (whichever land that may be) is a considerably higher aspiration than following in the footsteps of Peary or Ripert.

After science's acceptance of Biot's report and the cosmic origin of meteorites in the early nineteenth century, you might think that some people would have begun searching for meteorites as an occupation or at least an avocation. Perhaps too little was still known or had been publicized. Whatever the reasons, it was not until the early twentieth century (in America, at least) that the first of the meteorite hunters emerged and influenced all meteorite hunters after him. His name was Harvey Nininger.

The story of Nininger is best told in works of his own, like *Find a Falling Star* and *Out of the Sky*. He died on March 1, 1986 (just 4½ months after the death of another great meteorite pioneer, Lincoln LaPaz), at the age of 99. Nininger, directly or indirectly, may have led to the finding of more meteorites than all the other meteorite hunters in the world combined. Nininger was responsible for bringing out of the wilderness such giants as the Goose Lake (Calilfornia) meteorite, which weighed more than a ton and which he transported with four heavy horses and wagon in a trip that took a day and a half. His personal collection of meteorites, which he sold in 1960 to Arizona State University, consisted of 1,300 meteorites (the third-largest meteorite collection in the world) derived from 700 falls and weighing a total of 7½ tons.

Even Harvey Nininger (whom T. R. LeMaire ever so aptly calls "peripatetic") could not walk far enough to locate all these meteorites personally. Walter Scott Houston explains that although Nininger was a very solitary man, he learned to pretend that he was an extrovert. Nininger would bring a meteorite to the bars in small towns and stir up conversation. Another great pioneer meteorite hunter, Oscar Monnig, had to learn to temper his impatience with people who had seen a fireball or knew where some possible meteorites were located: he and his fellow hunter Robert Brown discovered they had to visit with these people, to actually sit down and spend some time with them, if they expected to receive the cooperation and information they wanted from the folks. Nininger seems to have been all over the Great Plains and western United States, talking as much as walking, in this way enlisting people's enthusiasm and help. Everywhere he went he planted interest from which meteorites seemed to sprout up, as if he had been a kind of meteorite Johnny Appleseed (even though his interest was not in disseminating meteorites but very definitely in gathering them). His work resulted in harvests for many later meteorite hunters. Walter Scott Houston asked his own students in Kansas to quiz their friends and relatives at home about meteorites, and some of them would return to college with stories of how their parents had met Nininger twenty or thirty years earlier and had later found suspicious stones they had saved in case Nininger ever passed back through. Some of the students brought the actual stones, which in some cases did turn out to be meteorites!

Walter Scott Houston is best known for the legendary "Deep-sky Wonders" column he has written for decades for *Sky & Telescope* magazine. But Houston is also an expert meteorite hunter, and how he developed his skill makes for an interesting and instructive story. As a high school student in the late 1920s, Houston could not find anything in print on meteorites, even in the *Encyclopedia Brittanica*. The local college geologist whom Houston consulted did not even know what a meteorite looked like. The cut and polished meteorites in a local museum were displayed to show only the insides—and the museum officials said that turning them around so that Houston could see the outsides was not "necessary." The young man persisted at museums until he often got his way, but in the end perhaps his greatest help came from his writing to and being answered by Harvey Nininger and Oscar Monnig. Monnig even sent a small sample piece of meteorite. Over the years, Houston tried to imprint on his mind the image of every meteorite or photograph of a meteorite he saw. Even still, it was not until he moved from Wisconsin (with its distracting glacial debris) to Kansas (a state famed above all others for its meteorite finds) that he began to have success. Within a year after the move, he found his first meteorite: the fifteen-pound Mayday stone, which he still keeps. In later years, Houston's trips to Mexico have yielded some especially good results, including an eighty-six-pound piece of the Xiquipilque iron, a meteorite known since before the time of Cortez.

In the sixty-plus years since Walter Scott Houston began his quest to learn about meteorites and find them, the amount of detailed information about them in print has increased but is still woefully inadequate. This is especially true of popular-level information and information about how to find and identify meteorites in the field. Perhaps it is not surprising, then, how few people are really adept at the field work. According to Houston, the number of people in America who are "really competent" at finding meteorites is about the same now as it was when he first became interested: no more than six or seven!

If you cannot meet any of these rare people, you can at least read some of their advice from time to time in the marvelous little quarterly publication of the American Meteor Society called *Meteor News* (only $3.95 for a year's subscription in the United States, $5.00 surface or $8.00 air mail rate elsewhere, obtainable from Route 3, Box 1062, Callahan, Florida 32011 USA). Along with many other excellent articles generally accessible to the beginner, *Meteor News* has in recent years published interviews with Houston, Oscar Monnig and Robert Brown, and other meteor and meteorite authorities (an audio tape of interviews with Houston and former Association of Lunar and Planetary Observers director Walter Haas is available for $8.00).

The two-part interview with Monnig and Brown (in *Meteor News* issues 67 and 68) is filled with excitement and humor as well as much information and advice. One of the best Monnig–Brown stories concerns the great Oklahoma–Arkansas fireball of June 8, 1920, which they were learning about from people in the course of another fireball investigation fully eighteen years later! People had not forgotten. As Robert Brown told John West and David Swann:

> . . . It was absolutely hard to conceive. It was a June night, fairly early in the evening, absolutely still, nothing moving and, what was real interesting to talk to some of the people, there were a lot of revivals. The corn was laid by, the farming was over and this an interim period. Mostly it's rural country over Eastern Oklahoma. And that cockeyed thing came by and lit up that country the way it had never been lit up. . . . It was much more brilliant than the sun. And one boy who pretty well had his wits with him said that place became absolutely, positively dead still. Animals . . . there was no noise whatsoever. And then this unearthly boom hit him. Literally just ripped things apart. That detonation was something to behold. There was another short period of silence

as the rumbles went back down the path. And he said then there was shouting and praying and everything you could hear all over that area.

Much more brilliant than the sun! Monnig and Brown mention that a weather observer at Fort Smith, Arkansas, who was probably at least several dozen miles from the fireball, wrote on his form: ". . . heat from the meteor was distinctly felt."

In the interview, Monnig and Brown allude to a case in which a shower of small meteorites may have rained on someone's roof but were not afterwards found—perhaps because they had been pecked up by chickens! Monnig says he did not have the heart to sacrifice one of the birds to "test that hypothesis."

Still another fascinating story was that of Monnig's quest to recover pieces of the Pena Blanca meteorite fall. At the Gage Ranch in Texas, a loud boom had been heard at lunchtime, and then the cook had run in. "Mr. Forker! Mr. Forker! I saw a big thing fall out of the sky and it looked like a sack of flour with the bottom coming out!" Two Mexicans soon after drove up to say that water had splashed out of a pool on the ranch, over a wall, and onto them. The Forkers did not want to waste water in that dry country, but they drained the pool a bit and found a 26-pound meteorite, "about the size of a big, round watermelon." Monnig and an associate had gotten word of the happening, and when they showed up and saw the 26-pounder they were dying to get out to explore that pool, as Monnig was convinced that more of the meteorite was out in the water. Instead of going out to get it, they had to sit through a six-course dinner, an evening of chatting about the stars with Mrs. Forker, and what must have been one of the longest nights of their lives. The two meteorite hunters were up early the next day and went out on their own, quickly gathering 10 pounds of fragments from the bank of the pool. After a big breakfast Mrs. Forker finally assented to go out with them and permit exploration of the pool itself. But then there was another problem: water moccasins. Even for a meteorite Monnig was hesitant about diving down among the snakes! Luckily, Mrs. Forker asked a ranch hand, "a big hefty guy with beautiful muscles in his back and arms," to do the diving. She showed him a piece of the meteorite and he agreed. Down he went and came back up in a while— with nothing. Monning urged him to try again. He went down and this time came up with . . a 106-pound meteorite! The water's buoyancy had helped him bring it up, but it was still a prodigious feat of human strength to get it to the shore, rear it up, and heave it onto the bank. Monnig wanted more: this was one of the rarest types of meteorites, and there was probably more in the pool. But Mrs. Forker gently declined. "So we had to rest content," said Monnig. "And we did." But even 40 years later: "There's some in the bottom of the pool yet, I'm convinced. And it's not weathered either, because it was a single-element stone. There was no iron. So it's still there in good shape."

What kind of men are these meteorite hunters? Men who are literally down to earth but also with a touch from something above—a bit of a cosmic perspective not common perhaps in the ordinary rock hunter. Men (and perhaps some women) who have one of the most esoteric hobbies— no, passions!—in the world (seeking to pick up the only things in the world that are, in the strongest sense possible, from out of the world). Yet, in order to pursue their solitary avocation, they are also men who must converse and fraternize with countless people from every walk of life. Perhaps we should not idealize them. Perhaps some of them have feet of clay all the way up to their necks. But consider the qualities that a great meteorite hunter must have: tremendous patience with others, with oneself, with nature, with life's limitations; the willingness to listen to others (while on the other hand not being facilely swayed from a quest); the wisdom to know when to go on (even against what less knowledgeable or less passionate people might think insurmountable odds)—as well as the wisdom to know when to give up. They are merely collectors of a kind of rubble, a rubble that most people do not care a whit or a jot about. But that is most people's loss. (What marvelous places this rubble comes from, what roots-of-our-selves significance it carries!) The best of what these men are reminds one of Yeats' "Fisherman." This was the man Yeats pictured in scorn of the controllers of the political and art scene, the pompous and insolent and phony people whose latest fashion in what they claimed we should believe and admire was as hollow as its predecessors. Such a man was simple and genuine, patiently committed to the simple things which are really the truest and most important. Yeats wanted to write this imagined figure a poem "maybe as cold and passionate as the dawn." This kind of man, not necessarily meek, is one who may inherit both the Earth and pieces of wonder from beyond the Earth—sometimes in the form of meteorites.

Sometimes people do not have to find meteorites—because meteorites find *them*. And when that happens it is generally in the most startling and unusual way.

Before 1982 only one fresh meteorite fall had been recovered in the small state of Connecticut besides the one at Weston that Thomas Jefferson had refused to believe: on April 8, 1971, a 2½-inch-wide, 12.3-ounce stony meteorite smashed through the roof and embedded itself in the ceiling of an occupied house in Wethersfield, Connecticut, a suburb of Hartford. So where and when did the next meteorite recovered freshly after its fall occur in Connecticut? It was a mere 11½ years later that a larger meteorite crashed through the roof and ceiling of another occupied house—in Wethersfield!

The second time was at about 9:17 P.M. on November 8, 1982, when everyone within hundreds of miles with a clear view toward Connecticut saw a brilliant greenish or orange-yellow fireball with a long yellow tail. In towns near Wethersfield the whole sky lit up, and afterwards loud, sharp reports were heard. Robert and Wanda Donahue were watching the TV show "M*A*S*H" when they heard a mighty thud that "sounded as though a truck had come through the front door." They ran from their family room to the living room and discovered a hole through roof and ceiling and what they thought was smoke—it was actually a cloud of plaster dust from the ceiling—filling the room. Furniture and rug were covered with pieces of plaster. Had this been a bomb? The Donahues called the police, and only minutes later one of the firemen who arrived found a six-pound grapefruit-sized rock underneath a table in another room and correctly guessed that this was the culprit and that it was a meteorite. (It had bounced off the floor, back up to the ceiling, and into the other room!) The Donahues were extremely helpful and permitted scientists to borrow the celestial visitor immediately. The Wethersfield meteorite thus reached a laboratory to be tested for its cosmic ray exposure more quickly than any in history—just two days after its fall. The meteorite went on tour, with stays at the Smithsonian Institute, the American Museum of Natural History, and Yale's Peabody Museum of Natural History, but the Donahues eventually decided that the Peabody was the best place for this Connecticut meteorite to permanently reside. To think it had hit a house little more than a mile from the 1971 meteorite is even more extraordinary when one considers the rarity of proven cases of buildings' being hit. In the 1961 book *Space Nomads*, Dr. Lincoln LaPaz and his daughter Jean stated that, based on the doctor's study, there had been since 1847 only ten documented cases of buildings' being struck by meteorites.

In the years since Wethersfield 1982 perhaps the strangest meteorite impact has been that of the grapefruit-sized one that smashed right into and stuck in a rural mailbox in Georgia—surely the most remarkable postal delivery (true "air mail") in all of history!

One of the ten meteorite strikes on buildings listed by Dr. Lincoln LaPaz is certainly the most notable of all encounters between a meteorite and an Earth-bound object—a human body!

The date was November 30, 1954, and the place the home of Mrs. Ann Hodges, a 32-year old housewife. She lived in Sylacauga, Talladega

(County), Alabama (a rather lumpily rolling volley of syllables). Her house, it turns out, was right across the street from the Comet Drive-In Theater! That day Mrs. Hodges was taking an after-lunch nap and so presumably was not among those who saw a flash visible in broad daylight across at least three states. But a few minutes later there was a tremendous noise. Suddenly a 7-inch by 5-inch, 8½-pound meteorite burst through the roof, hitting and damaging the top of the large console-type radio standing across the room from her and then bouncing in an arc of about 6 feet in maximum height to strike the wakening woman. Mrs. Hodges' first awareness was of pain on the left hip, arm, and hand, which the meteorite had bruised even through two heavy quilts, even on the rebound from the radio. A 4-pound fragment of the original body crashed and was found almost 2 miles away from Mrs. Hodges' house. Jets began searching for debris from what they thought had been a plane crash, and newspapermen were soon swarming over the Hodges residence. When the victim's husband, E. Hulitt Hodges, came home he was furious to find that authorities had already taken away "his" meteorite. Poor Ann Hodges was hospitalized the next day for the bruises and the shock.

Several recent studies suggest that meteorite impacts of buildings and people should be more common than we previously thought. About sixteen buildings a year in the world should receive damage from meteorites (obviously the cause of the damage must seldom be recognized by the owners). About once every nine years one of Earth's now 5 billion people should be hit. Where are the victims of meteorite impact other than Mrs. Hodges (and even she was hit only on the bounce)? Surely some must have been in the less technically advanced regions of Earth and were never brought to the attention of the world at large. There have been some claims of people hit, throughout history up to the present. We do know that a horse and a small dog were struck and killed by meteorites in the nineteenth century. Despite all that, however, Mrs. Ann Hodges of Sylacauga, Alabama, so far remains unique: the only person in history fully documented to have been hit by a "shooting star"!

Because meteorites are generally far less altered samples of early solar system material than Earth rocks, their finding and chemical analysis provides us with information on the very origins of the solar system and ourselves. If a meteorite is found soon after it falls, scientists can measure certain isotopes caused by the meteorite's exposure to cosmic rays in outer space—and thus date how long the meteoroid was an independent object. The evidence so far suggests that most meteorites are fragments of larger asteroidal bodies from which they were broken usually just a few tens of millions of years ago (though very much more recently and very much longer ago are also possibilities). The chemical structure of meteorites also seems consistent with their having been formed not all within a single major planet (somehow demolished) but instead within a dozen or so large asteroids that were the parents of all of today's thousands.

The chemical and structural classification of meteorites is complex, consisting of dozens of subtypes. Here we will pass over this large topic with just a mention that the most unaltered and significant type is probably the *carbonaceous chondrites;* the most mysterious type (if they *are* meteorites) the glassy *tektites;* the most beautiful features of many meteorites' outsides the often only one-millimeter-thick *fusion crust* and *"thumbprint"* indentations; the most beautiful features of many meteorites' insides the intricate, almost abstract *Widmanstätten structure* of intersecting plates of several different minerals.

In this romance of meteorites, let us conclude with a tracking of meteors that, by its precision, produces not only the end products (meteorites) that can be chemically analyzed but also the orbits of the beginning objects (meteoroids)—a revelation of these bodies' lives that is so much more complete than otherwise possible that it illuminates the whole mystery of what they are and how our solar system came to be. The ultimate technique for tracking meteors this well (radar and infrared detection aside) *should be* simultaneous photography of fireballs from different stations in a widespread network.

The *should be* deserves its italics. Even with numerous wide-field cameras with sensitive film in a vast network, the number of meteorite-dropping fireballs adequately photographed is few. Two of the three most famous networks were discontinued by their funders when their production of fresh meteorite finds did not live up to expectations. The other famous network had a failure in the mid-1980s that placed greater pressure on it, too. This surviving but endangered member of the famous three is the European Network, which expanded from a purely Czechoslovakian network soon after that itself got its start around 1964. In the same year, the Prairie Network of the United States was initiated, both it and the Czech network having been inspired by the successful photography of the Pribram meteorite's fireball in 1959. Although networks other than the European are still operating (for instance, in the U.K. and the U.S.S.R.), the Prairie Network was shut down many years ago, as was more recently the other of the famous three, the Canadians' Meteorite Observation and Recovery Project (MORP), which had become effectively operational in 1971. What set these three networks apart from others was not just that they were first, and large, and ambitious, but that they (if we count the Czechs' earlier success with

Pribram) were the ones that did produce the ultimate results: a total of three precisely determined meteor flights from which meteorites were recovered and accurate heliocentric (sun-centered) orbits of the meteoroids were calculated.

The first of these objects was Pribram, which fell near the city of that name in Czechoslovakia on April 7, 1959. Its fireball was magnitude − 19, about halfway in brightness between the sun and the full moon. The fireball was photographed by 11 cameras with the trails of about 17 pieces visible, though what pieces were found were located before calculations based on the photographs were made. Some of the early figuring suggested that the initial mass could have been very great, but more recent estimates have lowered it. The same is true for the terminal mass. Zdenek Ceplecha argued that in addition to the 5 fragments amounting to 13 pounds that were recovered, a piece weighing more than 200 pounds should exist. Douglas ReVelle calculated that a mass on the order of about 45 pounds was more likely. Although these quantities remain rather uncertain, the trajectory was accurately determined, and much else could be estimated with considerable precision. And the orbit of Pribram was most interesting of all. Its perihelion was not far outside the orbit of Venus, and, as expected for meteorites but never until then proven, its aphelion was out in the realm of the asteroids between Mars and Jupiter.

The next great success was that of the Prairie Network in the United States: the Lost City meteorite. On July 3, 1970, the fireball, brighter than the full moon, passed over northeastern Oklahoma, causing sonic booms heard across the 60 miles from Tulsa to Tahlequah. Because of severe winter weather, Prairie Network manager Gunther Schwarz took 6 days to drive to the vicinity of Lost City, Oklahoma. Films from two of the stations had been flown back to the Smithsonian Astrophysical Observatory in Cambridge, Massachusetts, and had been studied to give the expected impact point in the foothills of the Ozarks near Lost City. When Schwarz arrived, 7 to 9 inches of fresh snow was on the ground, and because several factors had not been favorable for the calculations, prospects seemed bleak. Just 18 hours after the fireball, a plane had been sent flying at 60,000 feet from Oklahoma City downwind to Atlanta and had succeeded in collecting more meteoritic dust than similar efforts had for the Revelstoke fall in Canada in 1965 and the important Allende meteorite fall in Mexico in 1969. As Schwarz started driving through the snow from Lost City toward the predicted impact point, it looked as though that dust might be all that would ever be found of the object. Schwarz was thinking about interviewing residents of the area—seemingly the only hope—when suddenly he had to put on his brakes: *there was a 22-pound meteorite larger than a football lying*

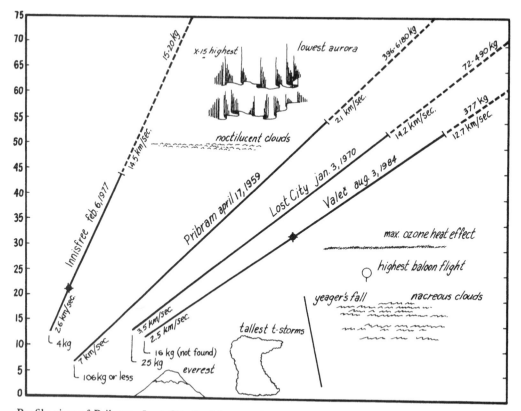

Profile view of Pribram, Lost City, Innisfree, and Valeč fireball flight paths. The mark at which a dashed line becomes solid indicates where the meteor first became luminous; the blacked-in circles with rays show the points of maximum luminosity (for Innisfree and Valeč); lower end of a meteor track's solid line indicates where it was last luminous. Both initial and final masses and velocities are given at their appropriate ends of the meteor tracks.

right in the middle of the snow-covered road! This was about a half-mile from the predicted spot, so a search of at least 200 to 300 acres working outward would have been required to find it.

The orbit and other statistics for Lost City were as accurate and as interesting as Pribram's. But (on a totally unscientific note) could any title sound more haunting than that of the Lost City meteorite? The real Lost City area, however interesting, must run a distant second to what the imagination conceives on the basis of the name. For meteorite enthusiasts it will always be the enchanted place frozen in time (or out of time) where that meteorite fell.

The third and last of the meteorites whose life was revealed by a photographic network was the one that fell on February 6 (February 5 local time), 1977. The Canadian MORP network photographed with two cameras the moon-bright fireball that produced a meteorite near the small town of Innisfree, about 140 kilometers (about 87 miles) east of Edmonton, in the province of Alberta.

The story of the Innisfree meteorite really begins with the mighty Bruderheim fall near Edmonton on March 4, 1960. Well over 600 pounds of meteorites were recovered from Bruderheim. This find helped inspire interest in developing a Canadian network, which was further flamed by falls with equally powerful and otherwise appropriate-sounding names occurring within 500 kilometers of Edmonton in the next 7 years: Peace River (1963), Revelstoke (1965), and Vilna (1967). The Innisfree fall occurred 10 years to the day and almost to the hour after that of Vilna. And by then the MORP network was waiting for it.

One observer near the fireball's path probably heard "anomalous sound" ("swishing"), and some people at other locations heard sonic booms ("a rumble, then a sharp rumble, mellowed rumbling for fifteen seconds, loud, then five seconds dying away," reported one observer not many miles from the town of Innisfree). The visual observations happened to be well enough placed to have led searchers into the right area, but calculations from the photographs permitted the search area to be narrowed greatly. Besides a tail wind to keep smaller fragments caught up with the larger, the steep angle of entry also contributed to making an unusually small "ellipse of fall." If meteorites were found in that ellipse it would be a "strewnfield."

The actual search began eleven days after the fall, at about 10:30 A.M. After about five and a half hours of cruising with four snowmobiles over the fields, scientist Ian Halliday spotted the major fragment. It was surprisingly small, measuring a little over five inches in length and weighing only about five pounds, but it was the real thing, and worth more than any gem to the scientists.

The orbit of Innisfree is shown in the accompanying diagram with those of Pribram, Lost City, and the Valeč fireball of August 3, 1984, whose meteorite the European Network mysteriously never found (despite a small—though heavily forested—search area). They are orbits very like those of Earth-crossing asteroids, as are most of the many dozens that have been calculated accurately for fireballs that did not drop recoverable meteorites but were adequately photographed by the networks. The number of fireballs that drop recoverable meteorites was found to be far less than expected (though perhaps Nininger-trained field workers would have found more), so despite the other tremendously interesting data gained, the people with the pocketbooks abandoned the photographic networks. David Meisel

Three-dimensional diagram showing orbits of Pribram, Lost City, Innisfree, and Valeč.
Generated by Guy Ottewell.

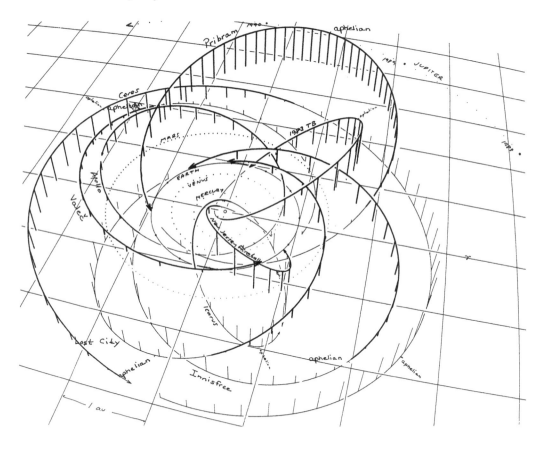

points out that the role of individuals like Peter Millman and Zdenek Ceplecha has been essential to keeping these worthy projects alive, and the retirement of such people can spell trouble for hopes of a network's continued existence (as perhaps Millman's retirement did for MORP). The ultimate hope is that new advocates will always spring up to keep networks alive or cause new ones to arise elsewhere so that the bright rain from outer space will never go unheeded and unexplored.

The romance of meteorites has no prospect of ending. More of the ancient iron giants will be found in the world's lonely places. More of the riveting fireballs will flare and thunder, and sometimes drop—even through the roofs of the unsuspecting—pieces of the solar system's puzzle, which we will, as lifetimes pass, begin to read with ever greater understanding and satisfaction. And meteorite hunters will keep looking, driven as if by deepest instinct to these pieces of otherworldliness vochsafed to our own—

I will arise and go now, for always night and day
I hear lake water lapping with low sounds by the shore;
While I stand on the roadway, or on the pavements gray,
I hear it in the heart's deep core.

William Butler Yeats
"The Lake Isle of Innisfree"

8

All the Worlds in
My Window

*Not a model, not a dream. All the family of the sun was
visible, in both simplicity and manifold beauty, in the
kitchen window of my childhood home that morning—
and not again in any window for a number of centuries.*

PEOPLE WHO FOLLOW THE PLANETS enjoy more than just the variety of
eight brothers and sisters of Earth, each constantly changing its appearance
and position. There are also those events called "planetary conjunctions,"
when one planet is near another or, technically, due north or south of
another. Every year brings with it some new examples of the endless possibilities of conjunctions and larger gatherings or sky-wide arrangements of
planets. The planets, observed with the naked eye or telescope, become
familiar (yet still mysterious) friends, and it is an unendingly varied source
of pleasure to have these friends posed in different constellations of the
zodiac and joining one another (as well as the moon and certain bright
zodiacal stars) in solemn, beautiful conclaves or merry, lovely meetings.

Yet the variations of the planets and their groupings is all the more
beautiful as a product of a mighty and complex theme of orbits which
underlies all. The longer you follow the planets, the more glimpses of this
order do you apprehend, and the more aware you become of the basic
recurrences: the nine-and-a-half month Evening Star, then nine-and-a-half
month Morning Star alternations of Venus; the one-zodiac-constellation-a-
year progress of Jupiter; the every-other-year "oppositions" and close approaches of Mars. If you watch these worlds long enough you even come
to experience a special eight-year cycle of Venus appearances, and Jupiter
returned to its zodiac starting point of a dozen years earlier, and Mars again

137

flaming the sky in its once-each-fifteen-or-seventeen-years closest oppositions of all.

How much richer it is to measure, evaluate, and treasure your life not
just in the dry numbers of mechanical clocks and paper calendars but also
in the multifarious moving lights in the sky and their interrelations.

The solar system is a clock with nine hands. The ends of those hands
are the planets, each farther from the sun, taking longer than its inward
neighbor to complete one full circuit or "hour"—and not only because it
has a longer distance to travel around its bigger circle: the farther a planet
from the sun, the slower it travels in its orbit. Thus Pluto (usually outermost), which has about 100 times farther to travel than Mercury (inner-
most), actually takes more than 1,000 times longer to complete an orbit
because it goes about 10 times slower. Most complexifying of all about this
"clock" : we live on the end of one of its moving hands.

The "times" you can read on the nine-handed clock of the planets are
always different, and you will never tire of them. This year may have a
dazzling Venus–Jupiter conjunction similar to one you recall from a few
years back, and by which you may remember much about what you were
doing, thinking, dreaming then. Yet it is only similar, not the same.

This essay is about episodes of sky glory surrounding the recurrence of
the largest significant and beautiful planetary arrangement of history—
which keeps happening, but not very often: barring tremendous develop-
ments in medical science, no one alive in the twentieth century will ever
see it again. The arrangement I speak of is the sonnet or bridge, the parliament or conclave, of all the planets in the sky at once.

Just once in about every 179 years, all the planets (excluding, sometimes,
tiny Pluto) are concentrated into a minimum span of their orbits. In other
words, the hands of the "clock" of the solar system are all pointing toward
almost the same "number"—almost the same constellation of the zodiac
composed of stars far out beyond our solar system. If we could stand on
the sun and look at such a time we would see all the planets, including
Earth, gathered in the same region of the zodiac. And that is what happened
on March 10, 1982.

You may or may not remember that date as one associated with an
especially regrettable episode in the history of pseudo-science called "the
Jupiter effect." It started with a 1974 book of that title by the British
scientists John Gribbin and Stephen Plagemann. Whether these men were
being foolish or were counting on the foolishness of a sensation-craving
public to make them rich by purchasing the book, we cannot say for certain.
Suffice it to say the book gave the impression that the March 10 concen-

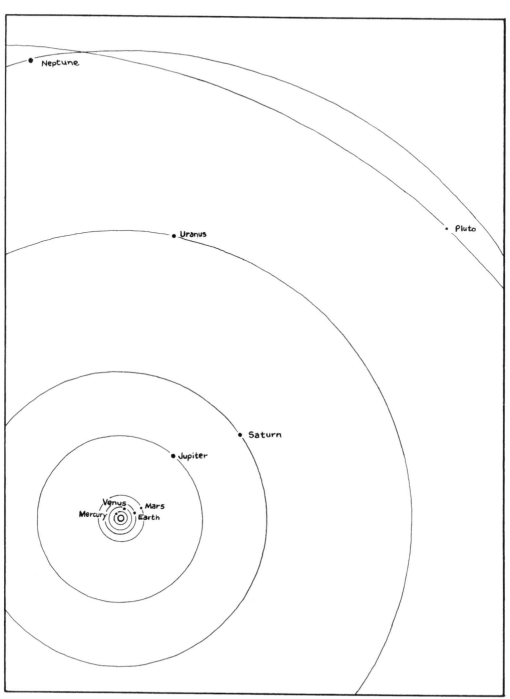

Minimum heliocentric span of planets, March 10, 1982.

tration would be so nearly a perfect line-up in space that all sorts of disasters would occur from the planets' tidal forces being all exerted along one line. In reality, the planets were gathered into only 95° of the full circle of the zodiac, just over 1 quadrant of the solar system at best. (In the clock analogy, all of the hands were not on one number but rather all nearly between the numbers "3" and "6" if we were looking from the north down on the solar system and measuring clockwise from the "First Point of Aries.") Also, even if the planets *had* been all concentrated into the same degree of ecliptic longitude, some would be a (relatively) little—actually, millions of miles!—north or south of the line that seemed to connect them so perfectly in two dimensions on paper. (I do not know for sure, but it may well be dynamically impossible for the planets to all lie at the same degree of ecliptic longitude at once—the four major satellites of Jupiter have orbital periods that stand in ratio to one another such that a perfect line-up of all four on one side of the planet can never occur.) Finally, even if the planets could all be in a perfect line (which they cannot), their combined tidal pull would raise our ocean tide only a tiny fraction of an inch! The two most important planetary pulls on Earth are those of Venus (closest) and Jupiter (most massive), and they combine their forces on us every four months—yet without disastrous incident.

So the popular misconception of Gribbin and Plagemann's idea—the misconception that the planets would *directly* tug our planet into earthquakes and other disruptions—was utter nonsense. But the two scientists had a different, more *elaborate*—but really little less foolish—mechanism by which the planets would *indirectly* cause cataclysms on Earth (including the great California earthquake—of course!).

Gribbin and Plagemann claimed that the planets would cause disruptions on the sun initially. Those disruptions would produce increased solar activity which would bombard the Earth with increased numbers of particles. These in turn would affect Earth's upper atmosphere and climate, and that—finally—would affect our world's rotation and produce catastrophic quakes. The only problem with the theory was that every one of its many cause-and-effect links was (and still is) unsupported by evidence (in fact, most of the links are actually opposed by considerable evidence).

If we needed any more proof that the "Jupiter effect" was not going to rattle our planet, we could (and those who believed *should*) have merely considered the record of history. There is no evidence of tremendous world-wide quakes and other cataclysms at intervals of about 179 years back from 1982. Even if we argued that Pluto, far smaller than even Earth's moon and so far out from sun and Earth, was somehow important to the "Jupiter effect," we could merely skip back to those recurrences of the 179-year cycle where it too was within the span. They were without incident.

Happily, while the "Jupiter effect" nonsense and scare came and went, genuine scientists were taking advantage of the planets' true and remarkable arrangement in a way which immensely enriched all our lives. They were bringing to us the results of what had once been planned as a "Grand Tour" of five planets and ended up as something almost as wonderful: the missions of the two *Voyager* spacecraft.

The key to these all-planet concentrations is getting the slow planets beyond Mars to all catch up to one another around roughly the same part of the same decade. (Mercury, Venus, and Earth race around their orbits to get into any concentrations rather quickly, and even Mars takes only a little more than 2 years to join the gathering.) As Guy Ottewell explained in *Astronomical Calendar 1982:* "After about 179 years, when Neptune has made just over one more circuit, Uranus just over 2 more, Saturn just over 6 more, Jupiter just over 15 more, Mars just over 95 more, and Earth just over 179 more, the situation of this year will be approximately repeated, though twisted about 40° onward around the heliocentric circle." Thus the last time the 8 major planets were concentrated into their minimum span *in the same part of the zodiac* as in 1982 was back in the fourth century A.D. (with Pluto not in the span). And the previous time before that would have been back (roughly) around the conjectural date of the Trojan War (with Pluto in almost exactly the same spot as in 1982!). But what was really important for space-mission planners about the 1982 concentration was the "grand curve" which preceded it.

The situation worked out (as it must with some but not all of the great concentrations) this way in the late 1970s: in their orbital race, Jupiter was still behind but catching up to Saturn, Saturn not far behind but catching up to Uranus, Uranus not far behind but catching up to Neptune. This curve of Jupiter–Saturn–Uranus–Neptune (as seen in the plan of the solar system) was just perfect for having spacecraft use each planet in succession to give a gravitational boost and speed it on to the next planet—far more quickly and cheaply than would ever otherwise be possible.

Only the American space program then had the wherewithal to rise to the challenge. The original proposal was to send two extremely well-equipped spacecraft, one past Jupiter–Uranus–Neptune and the other past Jupiter–Saturn–Pluto. But budget cuts reduced the plans to use of the pair of *Voyager* spacecraft, which were both sent by Jupiter (in 1979) and Saturn (in 1980 and 1981) and one of them, *Voyager 2*, on to Uranus (1986) and Neptune (1989). *Pioneer 10* and *11* were precursors to the *Voyagers*, *Pioneer 10* going past Jupiter in 1973 before being boosted onto a course that will take it out of the solar system while *Pioneer 11* encountered Jupiter in 1974 and was boosted to a pass of Saturn in 1979 before also beginning an exit from the solar system. These two *Pioneers* "blazed the trail" but returned far less detailed images and far less other data than the *Voyagers*, which revealed the two largest planets of our solar system and their rings and throngs of moons in an unprecedented flood of planetary discovery and beauty. Many scientists had been conservative about the chances of *Voyager 2's* power and critical functions surviving until it reached even Uranus, and *budget cutters almost took away the chance to receive the precious data even after the Jupiter and Saturn successes and after it had become clear that the spacecraft would almost certainly reach Uranus in operational order.* But in January 1986 a healthy *Voyager 2* became the first spacecraft ever to pass near Uranus—a planet discovered only just over 200 years earlier—and the revelations it radioed back were again even more than had been hoped for. At the time of this writing *Voyager 2* continues operating efficiently, and there is a strong probability that it will complete its truly epic mission by passing closest of all to Neptune in August 1989 before pursuing the same fate as *Pioneer 10*, *Pioneer 11*, and its *Voyager* sibling by leaving the solar system.

The thrill that Galileo had in being one of the first to look at the heavens through a telescope sometimes seems unimaginable; there seems no experience of our own to compare with that kind of astounding revelation. But this is not true! No one can live to see everything, but for lovers of the planets, our times (the past twelve years and, I hope, the next twelve or more to come) are difficult to exceed in excitement even by those of Galileo. I would go even further and say that if we are talking about the planets, then perhaps no time ever has or ever will bring so great a flood of revelations as the years around 1980. At that time the *Viking* discoveries on Mars were still fresh (still being made), Venus was receiving its most dramatic unveiling, and though our first close-up looks at one hemisphere of Mercury were many years past with the Uranus and Neptune encounters many years ahead, it was in the years 1979–81 that most of the incredible wonders of the Jupiter and Saturn systems were sent us. The "grand curve" and concentration of the planets came at just about the right time for mankind to be ready with spacecraft. If it had happened even five years earlier it would probably have been too soon for us—and then we would have had to wait until the middle of the twenty-second century for the curve to form again! The late 1960s, when the *Pioneers* and *Voyagers* were first being planned and budgeted, were also the highwater of excitement and support for the American manned space program. If the "grand curve" had formed even five years later than it did (certainly if ten years later—the late 1980s), I am doubtful that the American space program budget would have permitted anything like the wonderful *Voyager* missions to go.

Much has been said about the amazing coincidence of the "grand curves" occurring just when we could manage some kind of grand tour, even the lesser one that was the *Voyagers'*. Far less has been remarked about the marvelous planetary sights available to naked-eye and telescopic observers in the years around 1980, and the actual minimum span of all planets in 1982. Some of these sights and events were products of the great planetary concentration, but others were not and added to its spectacle wonders which were not available at earlier recurrences of the concentration in its 179-year cycle. I will therefore make a bold claim. I think this period from late 1979 to early 1984 may have been the most interesting—at least one of the few most interesting—in all of history for observing the planets!

After the preamble of the shameful "Jupiter effect" and the glorious *Voyager* flights along the "grand curve," I come at last to the aspect of the planetary concentration that is very close and dear to many a backyard observer's heart.

Let me first emphasize that to call any period "the greatest years for observing the planets" must to some extent be subjective, depending as it does so much on what kinds of sights and events one considers most important or enjoyable. No period will ever have the best of everything. Compared with the years 1979–82 or 1979–84, the years ahead may have more occultations (hidings) of bright stars and planets by the moon, or more extremely close conjunctions, or more of other kinds of planetary events which will delight us. Part of the fun is finding out what special things the planets have to offer that is best in a certain year or decade of your life.

So there is no need to worry that you missed all of the best there was if you were not watching the planets in that period in the early 1980s. But I still can say that to my way of thinking it may have been the all-around best, whose like will not occur again for many centuries or longer!

My first point about those years was their *paired oppositions*—by which I mean two planets reaching "opposition" (opposite the sun, biggest, brightest, most visible) within just hours or a few days of each other. Such instances are better than just having two planets at their best simultaneously, because the two will also be near each other in the sky for many months and usually engage in not one but a series of three official conjunctions, one planet due north or south of the other three times. This dancing back and forth to either side of each other is due to Earth's motion as it overtakes the other slower, farther-out worlds (only these farther-out worlds can come to opposition with the sun as seen from Earth). When one planet is over-taken by Earth we see it appear to move backward or "retrograde" to its usual direction of motion (the usual being gradually eastward) in front of the background of the distant stars. When there are paired oppositions (very rare), Earth is overtaking two outer planets, and there is double back-and-forth motion of them, together, in the sky.

The two farther-out or, more properly, "superior" planets that can become brightest are Jupiter and Mars. In 1980 the two were at opposition only 12 hours apart! They were near each other in Leo for more than 6 months and had three conjunctions with each other. Now it so happens that the orbits of Earth, Mars, and Jupiter are such as to prevent the latter two from doing anything even remotely like this more than just once every 143 years! The usual interval between two Mars–Jupiter conjunctions is more than 2 years: only at once-in-143 years intervals is anything better seen—the three conjunctions in about half a year. No other Mars–Jupiter conjunction except the central of the three ever (in our age, at least) occurs with the planets much more than about 90° from the sun and thus anywhere near their brightest and biggest. The middle conjunction of the 1979–80 triad had them meeting less than a week after their oppositions (when they are about 180°, the farthest possible, from the sun in the sky, very bright,

Retrograde loops of Mars, Jupiter and Saturn in 1979–80.

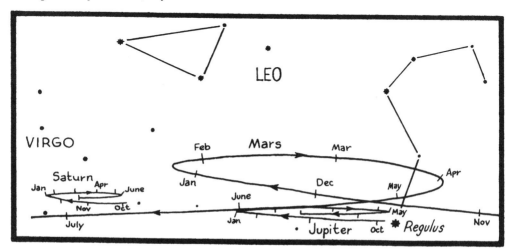

big in the telescope, visible all night long). I remember how splendid the pair were on the coldest night (down to $-23°F$. in Binghamton, New York) I have ever experienced, the Leap Day evening when they stood right beside the precisely full moon—and proved the only night-sky sight I have ever seen to actually exceed the sheer power and impact of the full moon! And yet the final of their three conjunctions was perhaps even more moving because it found them, still very bright, forming an extremely compact trio of gold, red, and blue with Leo the Lion's heart-star Regulus (a star they had been playing all around for those many months before this climax).

The historically most famous "Triple Conjunction" (series of three conjunctions) between two planets at paired opposition is that of Jupiter and Saturn. It happened in 7 B.C. when some scholars think it may have been what the Magi saw and became known as the Star of Bethlehem (lately the theory that conjunctions of Venus and Jupiter in 3 and/or 2 B.C. were the Star have come to the fore—if we seek a natural cause, I personally favor one of these). Johannes Kepler pushed the idea that the Triple Conjunction of Jupiter and Saturn in 7 B.C. was the Star of Bethlehem, himself inspired by a single conjunction of the two planets that he observed just before Christmas in A.D. 1603. It was Kepler, in fact, who first calculated that a Triple Conjunction had occurred at that date promisingly near Christ's birth (though the idea that a conjunction of Jupiter and Saturn was the Star seems to have been hinted at several centuries before Kepler). In any case, a conjunction of Jupiter and Saturn, the two giants of the planets, takes place always at intervals of about 20 years, but the Triple Conjunction much less often. Just how often is fairly complex. Different authorities give the average interval between Triple Conjunctions as 125 or 139 years, though much smaller periods between two occurrences are certainly possible—in the twentieth century, for instance, Triple Conjunctions of Jupiter and Saturn happened around the paired oppositions in both 1940 and 1981. 1981! Yes, in the year after Mars and Jupiter had their once-in-143-years Triple Conjunction, Jupiter and Saturn had their Triple Conjunction which takes place on an average of only about once in 1⅓ centuries!

According to Robert C. Victor, a Triple Conjunction of Jupiter and Saturn will occur only if the two reach opposition within about forty hours of each other. In 1981 that condition was easily met, with Saturn coming to opposition on March 27 only twenty-three hours after Jupiter (which had been at opposition on March 26). The two planets were within a few degrees of each other for even longer than Mars and Jupiter had been, in fact for most of a year. The appearance of Jupiter and Saturn in their Triple Conjunction is perhaps not so great an improvement over their usual single conjunction as is the case with Mars and Jupiter, but single conjunctions of Mars and Jupiter occur about every other year—even *single* conjunctions of Jupiter and Saturn happen only about once every two decades. The pairing of the two giants in their year-long stately dance was more majestic than perhaps any other conjunction of two planets could be. In 1980 much faster, closer Mars had pulled quite a distance away from Jupiter on its retrograde loop, larger and more northerly than Jupiter's—though the third, final, climactic pass of Jupiter by Mars had been close and occurred right by Regulus. In 1981 Jupiter and Saturn, both slow and with relatively small retrograde loops, had stayed together in close proximity. It was as if a great new double star, brighter than any other (though far less bright than the Jupiter–Mars middle conjunction near opposition), was adorning the heavens for most of a year.

To compare the Mars–Jupiter encounter of 1980 and Jupiter–Saturn of 1981 is to see that both were glorious, and in some considerably different ways. One final superiority of the 1980 event was the presence of a third planet not far off all year—Saturn, of course. It was only about a half-year's Jupiter journey ahead of (to the east of) Jupiter when Mars zoomed past Jupiter and Regulus in early May 1980. In June 1980, Mars shot past Saturn, but by 1981, when Jupiter caught up to Saturn, the speedy Mars was already away on the opposite side of the heavens. A compensation of great beauty in 1981, though, was the passing of Jupiter–Saturn very closely by Venus in August, when they formed an observable "trio" for the last time until 2080 (the previous time had been only ten months earlier as Jupiter approached Saturn!). "Trio" is a term coined by the Belgian astronomical calculator Jean Meeus and refers to any grouping of three planets (or two planets and a bright star, etc.) within a circle less than 5° in diameter in the sky. From August 24 to 26, 1981, Venus, Jupiter, and Saturn fit within a circle less than 3° across—during which time *Voyager 2* passed Saturn! A few nights later Venus had moved on, but the crescent moon joined the gathering to form a remarkable horizontal line of (from right to left) moon, Saturn, Jupiter, Venus, and the bright star Spica around nightfall as seen from the eastern United States (I recall catching only glimpses of this rather low spectacle through patchy clouds). The *trio* of Jupiter, Mars, and Regulus had been about twice as compact as the 1981 one of Venus, Jupiter, and Saturn, and though not quite as bright visible much higher and for much longer in the evening sky. Incidentally, the years just ahead—those of the early 1990s—will be especially superb for *trios*. According to Steve Albers, on June 18, 1991, Venus, Jupiter and Mars will lie within a circle as small as 1.8° with about the largest elongation from the sun possible for Venus (46°).

Planetary "Trios," 1988–98

Date	Objects	Smallest Circle	Faintest Object	Elongation from Sun
1988, Feb. 22[a]	Mars–Uranus–Saturn	1.36°	6.0	63°
1989, Jan. 12	Venus–Uranus–Saturn	4.61°	6.1	18°
1991, June 18[a]	Venus–Jupiter–Mars	1.80°	1.9	46°
1992, Jan. 11	Mercury–Ceres–Mars[b]	0.71°	8.5	19°
1992, Feb. 29[a]	Venus–Saturn–Mars	4.46°	1.6	29°
1995, Nov. 19[a]	Venus–Jupiter–Mars	2.04°	1.5	24°
1997, Feb. 12	Mercury–Uranus–Jupiter	1.05°	6.2	19°
1997, Dec. 24	Venus–Uranus–Mars	3.03°	6.2	32°

[a]Easily observable (the others are difficult or impossible).
[b]Ceres is an asteroid or minor planet, not a planet.
Data calculated by Steve Albers.
For trios through 2030, go to http://laps.fsl.noaa.gov/cgi/albers.homepage.cgi.

So Saturn was not far from the Mars–Jupiter pair in 1980, and the Jupiter–Saturn pair held the stage usually alone by Virgo's head in 1981 while Mars had fled off to the opoposite side of the heavens. But what happened in 1982?

We came very close to having a third year of a Triple Conjunction—this time Mars and Saturn! They were at opposition only ten days apart. Mars had almost caught the slowly retrograding Saturn for a first of three conjunctions when it stopped and retrograded itself—so there was only one official conjunction, which occurred later in the year when Mars came rushing back with direct motion. Despite the technical failure, though, the actual sight in the sky was almost as beautiful as it would have been in a Triple Conjunction, especially with Jupiter relatively near to the east of the two planets. For half of 1982, the constellation Virgo was bedecked with Jupiter at her feet, Saturn at her breast, and Mars lingering by her head but then gliding down the entire starry length of her past its brother planets.

The three truly great Triple Conjunctions possible almost occurred in three consecutive years, despite the fact that any one of them is typically an event which happens only once in far longer than a century. The only greater combination of their beauties might be if all three came to opposition on almost the same day in the same year so that there would be a triple Triple—nine conjunctions in little more than half a year, each planet meeting three times each with the other two. But I wonder if it is not better to have the glories spread out over a little longer—the triple Triple would be over too quick.

For any Triple Conjunction of planets to happen around the time of the once-in-179-years concentration is by no means necessary. To think that all these cycles came together in our time to produce such great technical and observational beauty is humbling.

Quite a few other rare planetary events happened during those years of the Triple Conjunctions and paired oppositions. For instance, Saturn's rings were seen edgewise from Earth three times between October 1979 and July 1980—the more typical triad of such an occurrence (it can also happen just one time in an occurrence), but any of these occurrences happens only once about every fifteen years. Yet this and even the Triple Conjunctions might be found auxiliary compared with two other events: two gatherings that provided opportunities to see all the planets in the sky at once—including in their smallest span of sky for the century.

The first event, occurring just before the time of the tightest heliocentric concentration, the "Jupiter effect" event of March 10, 1982, was *not* the tightest concentration of Earth's eight fellow planets *as seen from Earth*. That distinction fell to the planetary set-up visible in January 1984.

The 1982 period of all-planets-at-once visibility was longer, stretching for a month or more, but like the 1984 event essentially dependent on the visibility of flighty, elusive Mercury just before morning twilight grew too bright. Remarkably, the "Big Four" asteroids (actually the first four discovered)—Ceres, Pallas, Vesta, and Juno—were also visible at that same hour on mornings in February and early March 1982.

These asteroids were not bright enough to see with the naked eye, which is the way I did most of my observing: I beheld all five of the bright, classical planets at once (something possible, though usually quite difficult to do, every few years) and also a sixth, Uranus, just barely bright enough under those conditions for my naked eye. When I say "at once" I really mean with the ends of the span at the corners of my eyes or (more comfortably) in rapid scans back and forth.

While I enjoyed this marvelous sight, however, a small number of very ambitious amateur astronomers around the world were working very hard to see something even more marvelous: all the planets, including dim Neptune and dimmest Pluto, plus the four asteroids, in the telescope in succession on one morning. To make one's achievement of this goal really impressive, one had to see Mercury, last to appear, and then zoom back to find faint, faint Pluto, one of the frontmost, even with twilight beginning to threaten. A few people succeeded even in this tough test, though generally fairly large and well-equipped telescopes (along with tremendous preparation and sometimes teamwork) were necessary. I did not prepare very well myself (not even setting up elsewhere than my partly tree-blocked home site), though one morning I did see all the planets but Pluto and two of the asteroids. I also tried one morning before Mercury came into view to see the globes of the five other brightest planets in the shortest amount of time possible. To do so required enough magnification to see Uranus and Mars as dots, and I also required that I should take a full (though quick) look, not just an unappreciative glance, at each planet, steady in the field of view. I tried once and with a little luck managed the observation in little more than one minute.

My best view of the bright planets in this great gathering occurred on February 26, as dawn was coming strong but before its white-gold had yet won over the dusky violet of most of the sky. Last to rise, tricky orange-white Mercury peeked over the tops of the pine trees; well to its upper right hung the lantern of yellow-white Venus at its greatest brilliancy that very morning; much farther right burned yellow Jupiter, second only to Venus in brightness; then, not much farther west (right) stood proud, solemn, gold Saturn with bright star Spica just below it. Finally, just a few degrees more west, nearly at its closest to Saturn in this year in which they almost Triple-conjuncted, Mars led this majestic procession with a ruddy flame brighter than any star visible at the time.

Diagram of all planets in the sky, February 12, 1982.

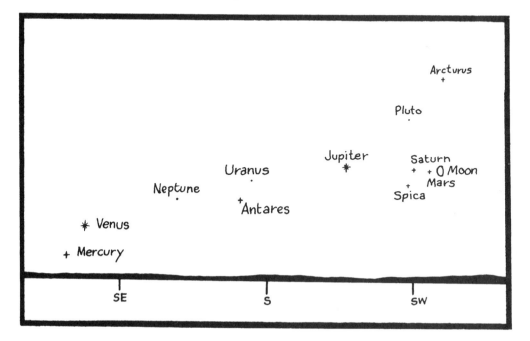

I thought of this entire all-worlds-at-once gathering as the sonnet of all planets. Somehow that beautiful, compact, strict rhyming verse form seemed an appropriate metaphor.

Most of us did not realize at the time of the 1982 gathering that there would be one more all-planet vision, in some important ways even better than the first. The last of our lives, it would take place less than two years later when Mars had gone all the way around the heavens and come back just before Jupiter could pass Neptune in the eastward direction around the zodiac and become an increasingly far-ahead frontrunner in the marathon of the slow gas giant planets.

From January 12 through 15, 1984, all eight of Earth's fellow planets were gathered into the smallest observable span of sky in the twentieth century. In the orbital race, Earth itself was too far behind to make the heliocentric span as small as it had been in March 1982. But the span of the other eight planets in Earth's sky was far smaller—less than 60°. That figure is in degrees of ecliptic longitude. So the eight planets were gathered into one-sixth of the zodiac's circle, the equivalent of about two average constellations of the zodiac. Because of some planets' being near constellation edges, the eight were actually distributed over four constellations: Mercury, Jupiter, and Neptune in Sagittarius; Venus and Uranus in Ophiuchus; Saturn in Libra; Mars and Pluto in Virgo—and no planet in the small wedge of Scorpius thrust up to the ecliptic between Uranus and Saturn. By the time Earth would catch up to this flock's rear edge in the race around their orbits, Mercury would be out and then back in, but Venus would be on the opposite side of its orbit, and thus out. But Earth's overtaking of the 6 superior planets for oppositions would occur in a period of only 10 weeks. And during January 1984 the great concentration would be internally further clumped into four two-planet conjunctions, two *trios,* and even a "quasi-conjunction" (two objects coming to within 5° of each other without either getting due north of the other for a true conjunction).

That January, there were the compact planetary *trios* of Mercury–Jupiter–Neptune and Venus–Jupiter–Neptune; there were five pairings between objects first magnitude or brighter less than 7° apart; there were three more pairings with a naked-eye planet and a telescopic planet less than 2° away from each other; there were three instances of a bright planet near a third magnitude star. The closest of these conjunctions was one of the closest of our lives, though involving a very dim planet and visible at best only in parts of the eastern hemisphere: Venus just 1.6 minutes of arc north of Neptune (only about 1/19th of the moon's apparent diameter)! The brightest of these conjunctions was one of the best of our lives between the two

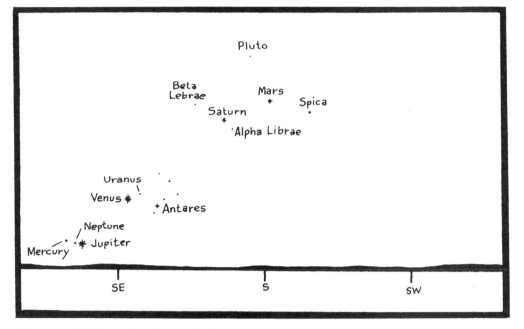

Diagram of all planets in the sky, Friday, January 13, 1984.

brightest planets, Venus and Jupiter, less than a degree apart on two American mornings—and joined by a very close crescent moon one American morning a few days later.

Unfortunately, for a few weeks in mid-January, clouds and cold limited observation of the great planet throng and its interior groupings in the eastern United States. At my home on the edge of the New Jersey Pine Barrens on January 22, I measured a low temperature of −17.5°F., equalling the lowest I had ever recorded there before. This time it was not mostly, as so often, from the strong radiational cooling of this forested area, but from an air mass really that cold—even Washington, D.C., measured −18°F. that morning, its lowest ever. A small town in Massachusetts set an unofficial all-time state low of −40°F. The full gathering of the planets was visible only in the almost coldest hour each morning, and few people in this part of America would have gotten much effective observing in if the temperatures had not greatly moderated from that January 22 nadir. But they did: the temperature was more than 73°F higher in four days, and more than 75°F higher in five days at my place!

So the temperature was in the 20s with clear skies on the morning of January 26—exactly twenty-three months after my most memorable morning with the earlier all-planet gathering. The planets were still not much more spread than 60° in ecliptic longitude, and in actual angular measure that meant confined to a span of little more than two Big Dipper–lengths! And this was one of the mornings of Venus's meeting with Jupiter.

Brilliant Venus was located directly above the not greatly less bright Jupiter: one mighty "star" piled almost right on top of the other as they lifted above the treeline to my east. Yet this was only part of a far larger arrangement whose beauty was all right before my face—not partly to either side—that morning.

Although I could not see it with the naked eye, I knew that Neptune was there, even closer to Venus than Jupiter was. To the upper right of the tight *trio,* the great constellation of Scorpius climbed with a fine moon and dim Uranus riding on his back. Still farther to the right or west, in the southern sky, two more bright planets shone together—not so closely paired as Venus and Jupiter, but much higher above the treeline and a lovely contrast of red and gold. This pair was Mars and Saturn, which a few weeks later were destined to have a very close conjunction. High above them, visible only in fairly large telescopes (though I once saw it in a six-inch telescope), was the feeble Pluto. As I stood there in strengthening twilight, I finally beheld the last planet to rise—fleet Mercury, not far to the left of the Venus–Jupiter–Neptune *trio.* Two years before, all the planets had been visible in one quick scan of most of the southern heavens from east to west. This morning, the ultimate dream of a planet watcher was come true: all of the planets condensed before me into one straightforward, nowhere peripheral view. It struck me like nine-branched lightning, held me still as a (deeply breathing) statue. All the planets at once.

The *Apollo 11* astronaut who did not walk on the moon, Michael Collins, was originally going to call his wonderful book on his career as an astronaut *The World in My Window,* referring to the sight of the entire planet Earth framed in his portal (in fact, hideable by his outstretched thumb when he was in lunar orbit). Instead, Collins changed the title to the also appropriate *Carrying the Fire.* That morning of January 26, 1984, I returned indoors for a while in mid-twilight and, in the still rather dark kitchen, stepped up to the south window. Even without putting my face very close to the glass I could see, glittering among the tree branches, the full bridge of all the planets and the moon. Several were too dim to detect with the naked eye, of course, but I knew that all the major worlds of the solar system—all the planets, all the moons (including Earth's)—were framed in my window. As a matter of fact, half of a kitchen window's view held the entire riches of the planets—Mars, Saturn, Pluto, Uranus, the moon, Neptune, Jupiter, Venus, Mercury, and the fair land and forests of Earth itself were all there—

a view not to be repeated in any window anywhere for more than a century and a half, or several times longer than that if Pluto were counted.

Only now, in looking over my notes, do I realize that at the start of the 1982 period of all-planets I got a sight that, strangely, seems to have been a prelude to the 1984 all-the-worlds-in-my-window observation.

It was on the morning of February 7, 1982, in the last hour of darkness, that I was awakened and wandered out into the dark house. It is difficult to explain, but I was in a state still touched by sleep yet somehow seemingly more awake than normal wakefulness. And then I saw visions outside three windows . . .

. . . Out the west window: the one-day-before-full moon, its giant white orb setting in the trees, silvering the branches.

. . . Out the east window: torch-like Venus hanging low opposite the moon and moon-window, it being the Morning Star newly arrived in the pre-dawn on its marvelous journey from a long, much-appreciated habitation in evening heavens.

. . . Out the south (the famous kitchen) window: red Mars, getting very bright now, rivaling the star Arcturus high above it, and followed closely by the pair Saturn and Spica (about equally bright) as well as less closely by the far brighter Jupiter—four jewels in pattern below Arcturus, riding the southern sea of sky halfway between the west moon and east Morning Star.

I went back to sleep and when I rose in daylight many hours later, this experience of the three windows seemed like a dream, yet with the perfect clarity of a vision—self-contained and itself glittering jewellike.

Today, looking back, it seems as though the experience of the three windows in the dark house in 1982 was the true prelude to the 1984 sight. It took two more years to get all those objects condensed into the one, south window. I would never see it again. Even my great-grandchildren's great-grandchildren would not be around to see it with Pluto included.

But to all of us these sights speak not a forlorn message of a greatest beauty, fleeting in a few mornings, now gone for so many generations. I say to you that if you go to three windows of your house (if you do not have the right house, then the one you should have!), go to them again and again in your life, you will see often out of at least one of them altogether splendid (and different) views of the planets, with maybe a bit of moon and stars thrown in for good measure. Similar but never the same will each view be. And perhaps you will grow so devoted to following these wandering worlds that you will get that especially magical window on the heavens, called a "telescope," through whose narrow portal you will behold seldom more than one planet at once but will marvel to rings and bands and icecaps and colors and moons more than you could imagine in a thousand nights of window-looking.

Then, whether or not you were pursuing astronomy in the early 1980s, whether or not you saw the things I observed, perhaps in your multitude of planetary sightings a glimpse may begin to grow. In the back of your mind you will suddenly begin grasping the motions of the great nine-handed clock, sunlit and glittering in the void, that is not merely mechanical but full of life-inspiring visions. And perhaps you will begin to see in the back of your mind one vision of all the planets, perfect like the lines of a celestial sonnet, like a troop of holy pilgrims, like a parliament of august lawmakers, like a bridge spanning emptiness, like the family they truly are—and like the gathering which just a few times among the passing centuries shines all together in the single window of whoever is then both alive and awake enough to see.

9

Making the Indoor Sky

SCULPTORS CAUSE the beauty of human bodies to dwell in living stone; painters build fires of form and pigment that burn with the beauties of humanity and the Earth.

But how to re-create the heavens?

It is interesting how difficult it is to make a realistic-looking re-creation of the starry sky. Anyone who has visited a good planetarium knows that their exceedingly expensive projectors can be used to portray some celestial objects and phenomena with quite enjoyable realism (though always far from perfectly, of course). Anyone who has visited a bad planetarium knows that they may manufacture little more than a monstrous parody of the heavens and celestial events. The failures in the latter case are sometimes attributable to the operators, special-effects designers and painters, or to the show writers (I find the role of the show-writer may be most important of all) but one thing is clear: good planetariums cost, and it is unlikely that an individual person or club can afford any such device that will not project a rather disappointing "sky." And you are unlikely to have frequent or easy—certainly not personal—access to a good one.

On the other hand, many star-lovers know the enjoyment of just looking at rich celestial maps and also of drawing sky-pictures. These things cost little or nothing and have the great virtue of being your own—especially your own if you make them yourself. Too bad that maps and pictures still leave something to be desired. How satisfying can stars be that cannot be seen in darkness, that do not shine?

There is a better way to make an indoor sky. Cost: a few dollars. Amount of work: always small in relation to the desire it satisfies. This method combines many of the best features of maps and drawings with those of planetariums. The result is a beautiful, educational, and amazingly realistic sky.

MAKING THE INDOOR SKY

The basic materials for this type of sky are heavy black construction paper and phosphorescent paint. Do not confuse phosphorescent with fluorescent paint. Fluorescent paint glows under exposure to ultraviolet light ("black light"). Phosphorescent paint glows on into darkness *after* exposure to ordinary light. Fluorescent paint will be easier to find at a local store, but somewhere in the area a supplier will probably stock the phosphorescent paint or can order it for you.

If you have only the most casual interest in astronomy or just want a novel decoration for your room, you can paint dots of random size and position on your paper and place it on your ceiling. But you can achieve many further degrees of authenticity, depending on how far you want to go.

The only extra resource you need for a very accurate sky is some good star charts. If you are an amateur astronomer, you probably have a star atlas. If you are a newcomer to the field, you can find star charts in a book from the library (free) or a book from a store (inexpensive). If you want to make an exceptionally accurate sky, you could also get a star catalogue, which lists the exact brightnesses of stars in the measure that astronomers call *magnitude*.

For an authentic sky, you should refer to two scales. One will set an area of the real sky equal to a certain area of paper for your artificial one. The phosphorescent heavens that I know are based on the convenient scale of 1 inch equals 1° of *declination* (the celestial version of latitude on maps of the Earth) and 5 minutes of *right ascension* (the celestial version of longitude on maps of the Earth). Thus a 12-inch by 10-inch piece of paper can hold one hour (60 minutes) of right ascension and 10° of declination. It is important to remember that the star maps you use will have a certain amount of distortion as a necessary part of their projection onto flat paper. You should not follow that distortion farther than it is intended. For instance, if you use the projection of the most popular star atlases, you may have to confine any one sky you make to between about +50° and −50° declination—and if you wish to do the regions around the poles, do them

The Indoor Sky:
Scale of Disk Diameter for Brightness
(5 magnitudes = factor of 10 in disk diameter)

Magnitude	Disk Diameter (mm)
0.0	32
0.2	29
0.4	26
0.6	24
0.8	22
1.0	20
1.2	18
1.4	17
1.6	15
1.8	13
2.0	12.5
2.2	11.5
2.4	10.5
2.6	9.5
2.8	8.7
3.0	7.9
3.2	7.2
3.4	6.6
3.6	6
3.8	5.5
4.0	5
4.2	4.6
4.4	4.2
4.6	3.8
4.8	3.5
5.0	3.2
5.2	3.0
5.5	2.5
6.0	2.0
6.5	1.6

separately (as they are done separately—with a different projection—in the atlases). This means slightly limiting the size of your indoor sky, but you have to do that anyway because of the fact that some parts of the ceiling will be much farther from you than others, wherever you are in the room. Actually, the 100° of declination from +50° to −50° is 8 feet, 4 inches on this scale, and in the other dimension you could fit two seasons of stars into 12 feet. That is enough to fill a small room's ceiling (and to represent most of the sky visible at one time outside). If you really wanted to make more, you could do all the seasonal constellations between the two polar regions in an east–west length of 24 feet—though this would be losing some realism (more seasons than you could really see in the true sky at once). Whatever size sky you do, make sure that your favorite constellation region is located most nearly over your chair or the head of your bed, where you can enjoy it best.

The other scale you need in making the indoor sky is one that relates the diameter of the filled circle of paint to the brightness of the star it is to represent. The accompanying table gives one possible set of values (any other scale will, of course, have figures that must bear the same proportions to one another). These suggested diameters are large enough to cause problems with accurately representing the brightest objects. If you make them any smaller, though, you will have difficulty seeing the fainter stars that really would be visible to the naked eye on a good night in the country. (If you are going to make an indoor sky, you might as well make it as star-filled as you could see on the best nights in the real one!) Of course, you could make smaller disks and compensate by super-charging them with a very bright, close light. But that is a lot of trouble, and the more faint stars will fade out of view all the same after a few minutes. With the size of stars given in the table, you will be able to see all but the most faint stars for hours after a modest charging by illumination from an ordinary light. Even just before dawn (or later in your well-darkened room), you could wake up in your bed to see the brighter stars shining over you.

What are the problems with representing the brightest celestial objects? The brightest star, Sirius, has a magnitude of about −1.4 (the lower the magnitude, remember, the brighter) and would have to be shown as a disk about 64 mm across. The even brighter planet Venus is sometimes magnitude −4.6, which would make it approximately 10 inches in diameter—totally unacceptable (it would practically fill up some constellations)! The best solution is to reduce the size of Sirius and the brightest planets. Sirius need only be 40 mm to look very dramatically brighter than a 0.0 magnitude (32 mm) star.

ADVANTAGES OF THE
PHOSPHORESCENT SKY

The most unrealistic aspect of this indoor sky is its representation of stars as large disks whereas, of course, they actually appear as radiating points of light of very much smaller size. But, surprisingly, every other aspect of this kind of artificial sky is so convincing that the overall effect is one of impressive realism. The soft, faint radiance of the numerous phosphorescent spots in complex patterns is unlike anything we would meet with in ordinary life—except the stars. You are in darkness that could be the darkness of a clearing in the woods at night, and above your head shine the stars, all glowing in their proper positions and brightnesses relative to one another. Even the apparent size of the constellations is right or nearly right—the eyes must actually be at a certain precise distance from any given area of the indoor sky for that area to fill the same angular field of view it would in the real sky. The scale of 1 inch equals 1° of declination and 5 minutes of right ascension works out pretty well in this respect for the height of most ceilings.

Although showing the stars as large disks is not realistic, several kinds of celestial objects can be portrayed with virtually perfect authenticity with this paint because disks do not have to be used or can be small enough to be unnoticed. I will turn to these types of objects in a moment.

But first, what about the value of an indoor sky in a lighted room? It makes an impressive mural. Far more important, it can be about the best celestial map imaginable. If you have spent some time making it accurate, it can show you at a glance the brightness of all conspicuous stars to within about 0.1 of a magnitude (which is about the best the eyes can do in nature). If you want, you can write in names and draw constellation lines (I prefer not to), making sure that they will not be visible in your artificial starlight. Or you can leave off names and lines and make your indoor heavens less of a standard (if giant and precise) map, more of a learning bridge between maps and the real sky.

OTHER CELESTIAL OBJECTS

An indoor sky is not a static thing because you can always make additions and because the movements of the planets (and other solar system bodies) can be enacted in it weekly or even nightly.

The inner planets Mercury and Venus move so quickly that charting them is a bit impractical (unless you are motivated enough to move them as much as several inches each day!). There are also problems with their brightness (Mercury's changes so often and so greatly; Venus is so bright that the only solution is to make its disk a bit larger than Sirius's and super-charge it with a bright, close lamp). Mars can also move swiftly in the sky at times, but it and all the other planets can be managed easily. Make them on separate pieces of black construction paper only slightly bigger than their disks, unstapling and re-stapling them when you change their positions (you can get their precise positions for several times a month in one of the popular astronomy magazines). Incidentally, I think the indoor sky is itself best put up with staples—if you have the kind of ceiling tiles which can take them.

To the naked eye, the planets differ little from stars (they do twinkle less), so they too can be made as plain disks of paint. But their disks can also be painted to resemble their telescopic appearance. Saturn can be given its rings, Jupiter its Great Red Spot and a few moons. To do this decreases the realism, but I personally find it easy to accept as a convention and well worth the sacrifice. Interestingly, with all the planets beyond Mars, their apparent size relative to one another (as seen in the telescope) does usually correspond rather well with the sizes their indoor sky disks must be drawn to have them at their proper brightnesses. (Remember, very bright Jupiter should be made considerably smaller than the formula would dictate.)

On the faint end of the scale, you can make objects with disks so small that their glow cannot be detected with the naked eye from down in your chair or bed. You can still see them if you get close up (with the lights on or off). In fact, if you have an optical instrument capable of focusing on objects so close as 7 or 8 feet away, you can even observe asteroids and faint planets (as well as small details on your brighter planets) just as if you were using binoculars or a telescope outdoors! Of course, there is a limit to how faint you can make things: when I tried to create a Pluto (magnitude about 13.7 in recent years) I could manage only a twelfth magnitude (con-siderably brighter) object with the use of a pin-point dipped in the paint . . . even when super-charged with light it could be seen with the naked eye only if you got your face to about a foot away from it! I know one person who has made a huge indoor sky with stars complete down to magnitude 7.75—thus many of them more faint than could be seen with the naked eye in nature by most people even on the absolute best of nights.

Star clusters, galaxies, and nebulae are three classes of objects that can be made with almost perfect realism. If you want, you can make them masterpieces of detailed painting—but your beautiful art will also be work-ing models, shining in your indoor sky.

Comets are another wonderful thing that can be portrayed with great authenticity. As a matter of fact, there may be no other way that the elusive glow of a comet can be faithfully depicted. Dilute your paint in varying degrees to render the different levels of brightness down to the most delicate, faint wisps.

The various areas of Milky Way could be painted with similar dilute mixtures, but perhaps it is more advisable to imitate reality and make a Milky Way out of thousands of very dim stars. I have seen a phosphorescent Milky Way that is a concentration of sixth to ninth magnitude stars, and it looks very genuine. (The painter made the fainter stars in great numbers by flicking paint off the bristles of a toothbrush!)

Some other celestial objects and phenomena take a little more doing. An aurora is a challenge. Perhaps the simplest way to make one is by shining a flashlight through a colored filter and off aluminum foil. Reflections onto your sky off the tiny random wrinkles in the foil produce interesting effects, especially when the foil is moved.

Good meteors could be made by painting just parts of freely hanging or swinging balls or other objects on string (when the painted part comes into view, you briefly see a streak of radiance glide across the stars). To produce a meteor shower or truly unpredictable behavior of meteors would probably require the making of an ingenious "meteor mobile."

You can use your sky to get glimpses you will otherwise never have of scenes too far away in space or time. What would the sky look like from one of Saturn's many, many moons? You can figure it out and then paint it to see for yourself in your sky. Would you like to see what Halley's Comet may have looked like at one of its closest and best returns? You can do it. How about re-creating a great conjunction or gathering of planets from hundreds of years ago—or getting a preview of one that will occur in your old age or after your death? What would it look like to see the already glorious Orion Nebula from ten times closer up? Paint an enlarged Orion Nebula based on your telescopic observations or on professional photographs of it in books.

There are fluorescent markers that make lines almost invisible in ordinary light but glowing in black light. One possible project is to use such markers (or similar paint) to make constellation figures (a lion for Leo, a crab for Cancer) and anything else you want (lines connecting the stars, labels, etc.). Then you should be able to flash on a good black light (it must not illuminate the darkness of your room enough to "drown out" the stars), and your figures will suddenly flash into view across the heavens!

Finally, the easiest special effect to make is a nova or supernova. Isolate a bright light on any object you want to "explode." The result is quite spectacular.

CONCLUSION

The phosphorescent sky costs little money to make, but it can be rich in realistic wonders. Twelve years ago, one amateur astronomer offered a friend of mine $100 for little more than an excellent winter sky (which probably cost little more than $5 to make). Of course, to create a very accurate indoor sky does take a tremendous amount of labor—the first time. After you have made your original, there are several ways you can turn out copies of it very much more quickly.

It is probably best to work on your first indoor sky only when the urge and the opportunity arise together, adding a little bit at a time. But you can go just as quickly or just as far as you want with it. If you desire only the brighter stars and do not care to have extreme accuracy in brightness or position, your work is much simpler.

And you will always have stars to sleep under.

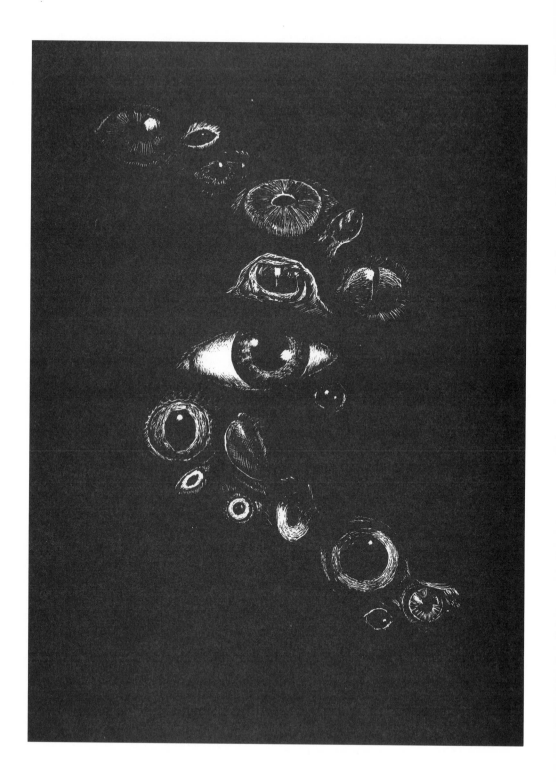

10

The Powers of Vision

SOME OF THE SECRETS of seeing are knowing where and when to look, and then looking well, with desire and wonder—often. As Thoreau said, "No method or discipline can supersede the necessity of being forever on the alert . . . the discipline of looking always at what is to be seen." Even so, while the most important part of *how* to look is with wonder and enthusiasm (and therefore frequently), there is also the *how* which involves the actual mechanics of vision. The purely physical powers of "ordinary" human vision are far-ranging, various, and mighty beyond what most people know. To observe the heavens well, and see many more of the marvels that are there, we must understand the physical powers, and admittedly the limitations, of that hero called the human eye.

Whatever optical instruments you are using—telescope, binoculars, or eye alone—the single most important ability for night observing is *light-gathering power*. The faintest stars the human eye can glimpse without optical aid are about 100,000,000,000,000 (100 trillion) times dimmer than the brightest light it can perceive without suffering damage. Even more amazing than this stupendous range, though, is how relatively easily the eye can adjust from seeing in one extreme of it, called day, to the other, called night. A landscape lit by the full moon seems much dimmer than a sunlit one, yet not enormously dimmer (as it really is). Artists and photographers have trouble conveying to us any difference between a moonlit and sunlit scene. But the full moon is actually about 465,000 times dimmer than the sun. And even blackest velvet in sunlight is several thousand times brighter than a white page of paper in the light of the full moon! These surprises are due to the eye's ease at adjusting from day to night lightning.

How can the eye function so well in conditions as different as—well, as different as day and night (which are even more different than we or the old cliché had imagined)?

The eye works with two fundamentally different (though partly overlapping) systems in bright light and dim, and every amateur astronomer or star-lover should know some things about the latter. Of course, in all vision light passes through the transparent outer covering of the eye, the cornea, and then through the black pupil to the lens, which focuses it on the light-sensitive retina at the back of the eye. The retinal cells are triggered by light to send electrical impulses which travel via the optic nerve to the brain. The differences in this procedure for bright-light and dim-light vision involve two things: the size of the pupils and the type of retinal cell predominantly used.

Everyone is familiar with the changes in pupil size due to changing light intensity: the pupils are small in bright light, large in dim. It is muscles in the colored part of the eye, the iris, which do the contracting and dilating of the pupil. The maximum pupil diameter achievable varies with individual and certainly with age, becoming less for a person with the passing years. Young and old alike have pupil diameters of roughly 2 mm or more in sunlight, but only the pupils of a person younger than about 30 years old are likely to dilate to as much as 8 mm (about one-third of an inch) in full darkness. The lens also grows somewhat less transparent with age, further reducing the eye's light-gathering power. The loss from these effects of aging, however, is not as important during most of life as observational experience and simply knowing how to use your night vision properly— good news for us all.

The other difference between how the eye works in day and night is the kind of photosensitive retinal cell which responds well in these different lighting conditions. Each of us has the two major kinds of photosensitive cell, which are known as *rods* and *cones*. Cones work well in bright light and can distinguish colors. Rods work well in dim light and cannot distinguish colors.

The rods work by virtue of a chemical called *rhodopsin*, or "visual purple." Bright light bleaches it out of the rods but it redevelops in darkness in a matter of minutes. Thus when you first go outdoors at night from a brightly lit interior you are not able to see faint stars for a while. Soon, however, as the visual purple is formed, your eyes become more and more sensitive to dim light, and the stars become more numerous as you glimpse ever fainter ones. Depending on how dark your environment is, how much your eyes were in bright light that day, and other factors, improvement of your ability to see faint objects may take about fifteen to thirty minutes to

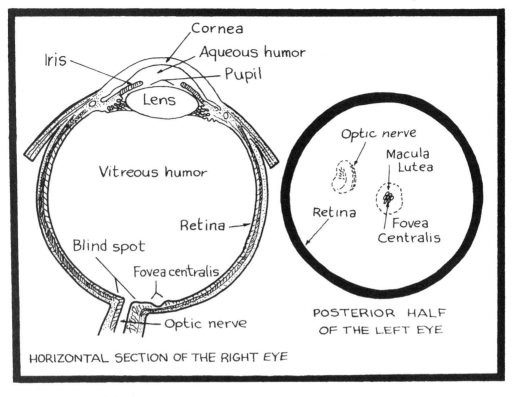

Structure of the human eye.

be mostly accomplished. A slight additional improvement of your light sensitivity can occur even after hours of being in the dark, though.

The process of the rods' becoming progressively more sensitive to dim light once they are in darkness is called "dark adaptation." Star-watchers should realize not only that the process takes a little while but also the importance of not interfering with it by looking at bright outdoor lights or flashlights. Shining a flashlight in the eyes of another observer is one of the cardinal sins, one of the most grievous blunders possible among serious amateur astronomers. Even looking at a star map illuminated by a weak flashlight is not ideal. Fortunately, the eye can recover very rapidly from a very brief exposure to bright light—if the light is not too bright. If you really want your eyes to do their best, however, consider using red light to look at your star maps. The rods of the eyes are hardly sensitive to very red light at all (the rods do not distinguish *any* colors as colors but very red light they do not even respond to as illumination). At the great annual meeting of amateur telescope makers and other amateur astronomers called

Stellafane (Latin for "the holy place of the stars") on Breezy Hill in Vermont, participants are advised to put red pieces of cellophane—Stellafane cellophane—over the light-emitting ends of their flashlights.

Such precautions help you see more of the richness of stars; however, they will not help much if there are already too many nearby streetlights or porchlights shining toward you or if the sky for miles around is illuminated by the lights of a city. The battle against this wasteful light pollution is one that all star-lovers should join (see Chapter 12).

Let us assume, however, that you live in a rural area or can travel to a site where man-made lighting is not a serious problem, and that the night is a clear one with numerous stars. Half an hour or more has passed, and your eyes are just about fully dark-adapted. What can you do to see faint objects better, and actually behold still fainter ones?

The most important technique, known to all experienced amateur astronomers, is *averted vision*. When you look directly at a very faint object, you cannot see it; direct your gaze a little bit away from it, and it becomes visible! The secret of this lies in the structure of the eye. Another difference between rods and cones is their areas of distribution on the retina. The cones are almost entirely concentrated in a small area called the *fovea centralis*. This is located at the center of your field of vision, and it is the only place where your vision is really sharp. Not only are almost all the cones in the fovea: none of the rods are located there. The result is that the center of our gaze is bad at seeing dim objects, and we cannot see dim objects very sharply. On the other hand, although the rods are distributed fairly generously across the rest of the retina, there is a narrow band at a slight distance from the fovea where they reach their greatest concentration and where the eye is therefore most sensitive of all to dim light. The band forms an oval which is located at about 19° to the sides of and about 15° above and below the fovea (the fovea itself is very small, accounting for only about 1½° of the eye's field of view!). In practice, of course, you do not really need to know precisely how far away from a faint star or comet to avert your gaze—merely knowing that you do need to avert it slightly and trying the technique frequently will set you well on the way to mastering the method. All people probably use averted vision a little bit, even if only unconsciously. Some authorities maintain that even a star of moderate naked-eye brightness—say about third or fourth magnitude—cannot be seen at all if stared at directly! The experiments to prove this are especially difficult to perform yourself once you have been consciously practicing averted vision for a long time—you find it hard *not* to use averted vision.

The faintest star you can see marks your "limiting magnitude" for the night. Of course, there is a limiting magnitude for your naked eye and for your eye with each of however many optical instruments you use. The interesting thing is that the commonly quoted limits are really quite conservative. According to an old formula, a telescope of six-inch aperture (six-inch-diameter primary mirror or lens) should show stars as faint as about magnitude 13.0. In reality, experienced observers often can perceive fourteenth-magnitude stars (that is, several times fainter) in clear, dark skies with such an instrument.

Just as conservative is the traditional naked-eye limit of magnitude 6.5. I remember our delight when a friend and I first tried using a limiting magnitude chart of various stars published with Walter Scott Houston's "Deep-sky Wonders" column in *Sky & Telescope* magazine. The field was around the Keystone pattern in the constellation Hercules, which passes overhead at 40° N latitude. When you attempt to see the very dimmest star possible, you should always look for it high in the heavens because the lower in the sky an object appears, the more it is dimmed by the longer pathway of atmosphere it has to travel through. (This dimming effect is called *atmospheric extinction* and is made all the worse by haze, dust, and man-made light and air pollution—consider how the sun is often tremendously dimmer when near the horizon.) Unfortunately, Hercules had long since passed the zenith that first evening my friend and I tried the limiting magnitude chart. But that was why we were even more cheered by finding that we could see a 6.6 magnitude star with relative ease. The fact is that many experienced observers at excellent sites regularly see stars of magnitude 7.0 and fainter on very good nights. Although it is difficult to test, my suspicion is that not even much experience and certainly not especially sensitive eyesight is needed for reaching such levels of faintness.

Why is there such a large discrepancy between the oft-quoted traditional limit and the real limit? I believe it is simply because a person must bring together several things and use them with considerable enthusiasm and patience to succeed. One thing needed is a good star atlas, and preferably also a good star catalogue that lists the exact magnitude of stars down to the sixth and seventh magnitudes. The second thing needed is a very clear night with little or no interference from light pollution—and admittedly the latter part of this requirement is getting harder for people to meet. Once you have these things and find a good field of stars that will be nearly overhead, you are ready. You should realize, however, that some real patience and effort are going to be required to go far past the traditional limits. Some of the observers who originally established these conservative limits may not have had the patience.

But is it after all worth the preparation, effort, and intense concentration to see somewhat fainter stars? When a comet comes along and you want

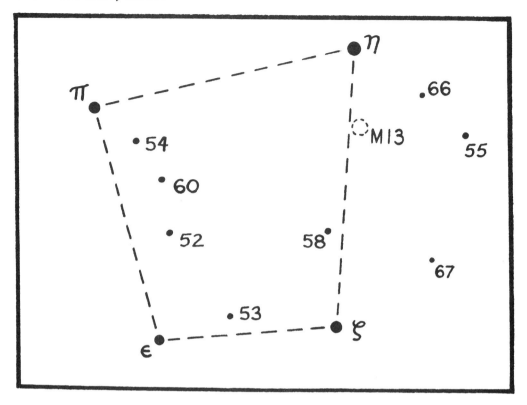

Test stars for limiting magnitude around Keystone pattern in Hercules (decimal points omitted between numerals in magnitude figures). Adapted from *Sky & Telescope*.

to perceive millions of ghostly miles of a tail that other observers do not even suspect, yes! There is also the very strong general argument that seeking your faintest stars will increase your powers of observation. This increase will reap benefits for you in everything you view, whether it be the rings of Saturn or the spiral arms of a galaxy or the faintest wisps of a nebula's outer boundaries. There is a less practical but to many of us no less compelling argument: it is the great pleasure in looking forward to not just clear nights but those few nights of the year—or one best in many years!—when your trained eye can prove that the atmosphere is the clearest of a thousand nights, the clearest you have ever seen, probably the clearest anyone you know has ever appreciated or understood. What a triumph when you can demonstrate to yourself that you are really seeing dimmer stars than ever before—that the night heavens stretching above you really must be the most richly star-clad you have ever beheld.

I can add several pointers to the basic ones of allowing dark-adaptation and using averted vision in your quest for faint stars. One is the fairly obvious advice that you should use your hands or arms to help block off most of the sky except the part your target star is in. This is not just a matter of eliminating distractions, because the combined radiance of the entire heavenly host and background light of the sky (caused by several sources) can actually affect your dark-adaptation slightly. Another important point is to rest and to close your eyes at least every few minutes. While you are actually trying to observe the very faint object you should also move your eyes at various intervals for several reasons. First of all, even bright objects kept in the same part of the retina are eventually ignored by the brain and "disappear" in a remarkable trait of the visual system called the "Troxler phenomenon." About ten times a second the eye almost irresistably (and of course quite involuntarily) travels to and fro in short arcs and helps prevent this phenomenon, but we can experience it to some degree with too long a period of deliberately fixing our gaze in one place. A second reason for moving the eyes—sometimes fairly rapidly—while seeking a faint object is to help notice it by its springing alternately in and out of visibility as that narrow band of most sensitive rod-rich retina passes across it. A final pointer that I have never seen mentioned but which is extremely important is helping to fix the position of a faint star you are seeking by using geometric patterns of brighter stars nearby. The human eye–brain system has a strong propensity for finding geometric patterns and especially lines in nature (the latter is probably not unrelated to one of the few instincts we are almost surely all born with—the fear of falling). Try to fit your target star on the star atlas into several lines or simple triangles with brighter stars, and you will find this an invaluable aid to your concentration.

On my best night ever, I believe I was glimpsing with the naked eye stars down to near magnitude 8.0! Such a sky starts becoming literally star-crowded, which actually becomes a problem. There are so many stars at this and slightly fainter magnitudes that neighboring stars can contribute their brightness to the glimmer you glimpsed and supposed was coming just from an individual object. You may have to seek out one of the more star-poor regions of the sky. You have to look through a good star atlas carefully for candidates, check neighboring stars with binoculars, and use extreme caution in accepting your observations until you are convinced of what you are seeing. Of course, now I am referring to the kind of effort which not all star-gazers will be willing to make, only those with a special taste for this pursuit. The world's increasing light pollution has been making this already rare activity all the more so. But some of us continue to have a fascination for the quest. We ask: how faint is the dimmest object ever perceived by man with the naked eye?

I am not aware of anyone who has seen significantly fainter than magnitude 8.0 from near sea level. My best observation was made at such an elevation, and I doubt whether it is often matched. And yet I am dubious that my eyes are any more sensitive to dim light than the average person's. More recently, a spectacular series of observations culminated in the visual recovery of Halley's Comet at the 1985–86 return. *Sky & Telescope* writer Steve O'Meara saw Halley's Comet with a twenty-four-inch telescope when its magnitude was only 19.6, and the faintest stars in the field were roughly a magnitude fainter: both probably records (for faintest comet and faintest object visually perceived) but presumably surpassable by a great observer like O'Meara who was determined enough and used a much larger telescope. O'Meara was observing near the 13,800-foot summit of Mauna Kea in Hawaii, so he had the advantage of having much of Earth's atmosphere below him. On the other hand, mountaintop observing has its drawbacks, including the fact that the reduced oxygen levels at high altitude reduce retinal sensitivity. O'Meara tried to compensate for this by judicious use of an oxygen mask. Still, the many problems he overcame are a tremendous testimony to his ability, and his story (recounted in the April 1985 issue of *Sky & Telescope*) is inspiring reading. All of which is prelude to saying that Stephen James O'Meara, on the night before his January 22, 1985, recovery of Halley, adjusting to the thin air first at the 9,000-foot level, was able to reach a limiting naked-eye magnitude of 8.4!

Has this ever been bettered? To the best of my knowledge, this is the faintest the naked eye has ever seen in nature without any aid. But there is aid other than the greater light-gathering power of binoculars and telescope. Back around the turn of the century, an experiment at Lick Observatory in California used a blackened tube through which a star whose position was precisely known could be sighted—this was, in other words, a kind of improved version of the cupping-of-hands-around-eyes technique that observers typically use anyway. The faintest star that could be identified in this experiment was magnitude 8.6.

Robert Burnham Jr. mentions laboratory experiments in which artificial light sources of as faint as magnitude $8\frac{1}{2}$ could be seen against a perfectly black background (which the night sky never is). So perhaps about magnitude $8\frac{1}{2}$ really is the absolute limit of the naked eye's light-gathering power. I am not so sure. O'Meara just sketched the dimmest stars he could see in a part of the constellation Puppis and much later checked their magnitudes. I do not think a person ever sees his or her very faintest without having a target object to strive for and knowing beforehand its exact position. O'Meara may therefore be capable of glimpsing considerably fainter objects.

   Although we now reach the limits—still uncertain—of unaided human vision, there is one more logical step. No, not the step into space—not yet. A variety of problems still beset astronauts who would use the complete absence of atmosphere between them and the stars to permit viewing of still fainter objects. Someday, perhaps soon, humans will overcome those problems and see to something like a half magnitude fainter than they do near sea level under Earth's atmosphere. That may not seem like much, but, as Isaac Asimov pointed out long ago, in this context the number of faint stars is so much greater than bright stars that we are talking about nearly a *doubling* of the total visible even if we improve our limiting magnitude from 6.5 only down to 6.9. The faintest stars are glimpsed only with skillful averted vision, of course, so the effect is really not quite the same as doubling the number of brighter, directly visible stars would be. Yet this immense increase in the number of stars is something to bear in mind when you consider whether improving your ability to perceive faint objects is really worth the necessary effort.

   The final step in our quest for the most light-sensitive eyes is not into space. It is elsewhere in the animal kingdom. What are the faintest stars visible to the most powerful light-gathering eye of any creature on Earth?

   Scientists have proved that many animals navigate by the stars (though apparently aided by other remarkable means, too). One experiment in a planetarium indicated that certain moths—which are, of course, nocturnal creatures—can perceive almost as many stars as humans can. Animals whose eyes reflect light back to us in the dark—dogs, cats, deer, bears, raccoons and others—do so with a layer behind the retina called the *tapetum lucidum*. Light that passes between the light-sensitive cells of the transparent retina on the way in is reflected back by the tapetum and thus gets a second chance of being perceived as it passes again (this time in an outward direction) through the retina. The second crossing causes some blurring, but such animals can gather at least several times as much light as we do. If, as one authority says, a cat's eye can gather six times more light than ours, then a cat should be able to see stars roughly two magnitudes fainter. No wonder cats are among the creatures whose pupils can close to narrow slits in bright sun to protect their eyes from being overwhelmed by light.

   But cats' eyes are not the greatest light gatherers. The champions, not surprisingly, are the eyes of owls. These are even larger than they look, filling most of the bird's head. The retina is very much richer in rods than the human eye's. Experiments suggest that the eyes of some owls collect at least *fifty* times as much light as ours. Most owls do not migrate, and we do not know—may never know—what the brain of an animal other than the human really "sees" (that is, what kind of image finally registers in the

brain). But imagine for a moment that you are an owl flying, on absolutely silent wings, through a clear, dark night. . . .

Six of the seven stars of the Big Dipper burn so brightly they rival the Venus and Jupiter of human vision—as do several dozen other stars. Even the faintest stars that most humans see on a good night are as bright as the North Star to you. You can perhaps distinguish by their fuzziness and size several hundred galaxies and star clusters that humans view only through telescopes, a few at a time. Wisps of nebula, some giant, are common. The faintest stars you can see are eleventh, twelfth, maybe even thirteenth magnitude. That is comparable to what is visible to humans through a telescope of three-inch or four-inch primary lens or mirror, but you are not beholding stars to this limit in a telescope's tiny field of view but rather across a vast part of the heavens' scope. Instead of seeing the few hundred stars in a telescope's field or the maximum of a few thousand that a human with unaided eyes can glimpse in the entire heavens, you fly soundlessly with head spinning around as fast as an eye's blink, and you are scanning perhaps a million stars in the sky at once. While most moon-sized sections of the sky hold no naked-eye stars for a human being, the average one contains several hundred for you! The only view of the heavens more glorious than yours is the one which humans just may conceivably get in some distant future: the view from a planet in the ball of possibly billions of stars that is the central hub of our Milky Way Galaxy.

Less can be done to facilitate the ability that, after light-gathering power, is the second most important of the human eye in astronomical observing. If we are talking about the powers of the eye itself (without aid), that ability is not of course magnification, for the naked eye's must always be 1X. It is rather visual acuity, sharpness of vision, which astronomers usually talk about in terms of *resolving power* or *resolution*—the ability to resolve fine details.

Quite a few people are gifted with natural vision better than the standard 20-20 (at least as good as 20-10), and eye doctors often seem to provide their nearsighted patients with glasses that correct vision to slightly better than this supposed norm. The differences in people's sharpness of vision are of course mostly attributable to how well their eyes can focus, which the eye does by the use of muscles that change its shape and the distance from lens to retina. Presumably some people do not have significantly more cones in their fovea than others. There certainly are animals with more cones in their retinas than humans, however. The champion again comes from among the birds, in this case some of the hawks found capable of seeing detail at distances which indicate that they have about eight times better resolution than a human's. We might make this more impressive by converting to magnification and saying that such a hawk is naturally equipped with (unbelievably wide-field) $8\times$ binoculars for eyes!

Even 20-20 vision is capable of feats of resolving that few people would suspect possible. Once again, the most important factor is to have enthusiasm and patience enough to try and to concentrate. What the eye can accomplish in bright lighting is amazing. The great explorer and scientist Alexander von Humboldt tells of seeing individual human figures from distances of many miles away across the valleys in the clear, thin air of the high Andes. He also cites experiments—for instance, one with visibility of spider-web strands—which demonstrate that the human eye can resolve features down to less than a second of arc (better than 1/1,800 of the Moon's apparent diameter!). Experienced astronomical observers will find such claims simply incredible because that is smaller than the apparent diameter of any of the planets except Pluto. People do not walk out and see with the naked eye the globe of Neptune—more than $2\frac{1}{2}$ billion miles away! To perceive Neptune distinctly as a disk requires a magnification on the order of $100\times$ or more with a good three-inch, four-inch, or larger telescope on a good night. There is no properly documented report of anyone seeing the *shape* of any of the planets with the naked eye—except possibly Venus, which can have an apparent diameter several dozen times larger than Neptune's. Are Humboldt's reports balderdash?

I do not think so. With glasses correcting my vision to about 20-20, I once tried a visual acuity test involving a line I drew in black ink on a white piece of cardboard. Taking it out on a clear, sunny day, I found that I could distinguish this line—just 1 mm wide—from as far as about sixty yards away! Why, then, do I not see some of the planets as tiny globes in the sky without optical aid?

The sightings of my line or of spider-web strands were achieved over relatively small distances through the atmosphere, and Humboldt's in the Andes through the very thin clear air at high altitudes. All of these observations also involved extremely strong contrast—no planet contrasts with the night sky as much as my black ink did with white cardboard. My experiment and the one with the spider web also benefited from the human eye's talent for perceiving lines, even extremely thin ones. The most outstanding and significant difference between these sightings and most astronomical observations is of course yet another factor: the sharpness of vision permitted by the cones in bright light as opposed to in dim. The sharpest vision, remember, is found in the fovea—but the fovea is composed entirely of cones, which function well only with rather bright light.

The only way to determine how sharply the eye can see in astronomical observations is to test it on certain objects. The problem with this is that the types of test objects and the conditions under which they are visible vary greatly. First of all, the turbulence of the atmosphere, the "seeing," is important: a person with extremely sharp vision may do more poorly in resolving on a night of poor "seeing" than a person with average eyesight on a good night. Through the telescope it can be appreciated that star images are in a sense thrown out of focus by atmospheric turbulence—they are not small and sharp but rather large and blurry on such a night of bad "seeing."

Unfortunately, it is not only the atmosphere but also certain irregularities of the eye (all eyes) that spread out the images of point or near-point sources of light. We do not see stars or (with the naked eye) planets as intense specks of light but rather with rays—the brighter the object, the longer its rays. What causes this admittedly beautiful distortion? Our pupils are not perfectly round, and so diffraction caused by the tiny irregularities on their edges is one cause. Almost certainly more important are distortions in the edges of our lenses caused by their attachment to muscles. An experiment which M. Minnaert suggested is to avoid the edge distortions by punching a hole a millimeter wide in a piece of cardboard and looking at a bright star or planet with this hole centered directly in front of your pupil. That is easier said than done, but once it is done you will find that the star or planet looks quite like a round spot—without rays. This might be an excellent technique for improving your resolution if it did not drastically reduce the amount of light getting to your eye, especially to many of the rods you need to use. The typical eye sees the ray-spread of a very bright planet on a dark night as several minutes of arc wide. But the brighter the sky from moon, twilight, or light pollution, the shorter the rays will be (your pupil contracts and begins to do what the cardboard did in the experiment—block off the object's light from falling on the distorted edges of the lens). So even in this effect of rays or "diffraction spikes" there is considerable variety according to conditions.

We know that the eye and atmosphere always fall short of perfection for tests of visual acuity in the heavens, but can we at least establish some standard situations so that different people's vision can be compared and a rough idea of our ultimate limits of acuity in the heavens can be obtained?

The problems continue when we consider the very many different and hard-to-compare tests offered by various objects in the sky. Take, for instance, "double stars." A double star is two stars extremely close together in the sky—the two may be suns that are actually traveling through space together or a mere chance line-of-sight arrangement from our point of view (one star actually much farther away than the other). Double stars would seem to be the ideal test of visual acuity: is your vision sharp enough to see the double star as not one but two separate points of light? The closest-together pair you could split would show your limit. Or so it might seem. In practice there are several factors that complicate the issue. First of all, there are not as many double stars visible to the naked eye and just splittable by the naked eye (thus a good test for average *vs.* excellent eyesight) as we might want. The solution of using telescopes adds the complications of different optical systems, in different states of "collimation" ("collimation" is alignment of the optical components intended to give a good image). There is for telescope users the standard called "Dawes' limit," which is supposed to give for any particular size of telescope aperture the best possible resolution (hence the closest double star pair you could split). But observers sometimes exceed Dawes' limit. And whether you use a telescope or the naked eye, there is a further complication with double stars: it is progressively more difficult to see the more faint of the two stars as the difference in the two's brightness is greater. The easiest double star to split is one in which the components have exactly the same magnitude. But that is unusual. Which is easier with the naked eye, if the separation of components is the same in both cases: a first- and third-magnitude pair or a fourth- and fourth-magnitude pair? Many such questions can be asked, and the answers all depend on several factors besides the particular conditions of the sky and the observer.

The situation is not as hopeless as it seems! We just need a lot of observers testing how well they can see various double stars on numerous nights—and making sure to note carefully the sky conditions. To that end I provide the accompanying list of famous double stars on which to test your naked eye. The toughest splits on the list are Steve Albers' separation of Mars and Epsilon Geminorum during a conjunction and the supposed splitting of Nu Draconis, which Patrick Moore mentioned but did not document (and is so nearly incredible as to make us cry out for supporting evidence). Although the Mars–Epsilon Geminorum event was extremely unusual in some ways, there are suitably close and well-timed conjunctions of planets with bright stars available to you every few years—be on the lookout for them in advance in the pages of astronomy magazines and astronomy almanacs. And what about that almost always available and widest split on our list, the pair at the bend in the Big Dipper's handle, Mizar and Alcor? The pair were popularly known as Horse and Rider long ago, and another name meant "The Test"—but surely a test of bad rather than excellent vision. Do not be upset if on a particular night you do not glimpse Alcor easily, or at all. On a good night (clear, dark, and steady) you should see it, though, for Mizar and Alcor were not called "The Test" because only sharp-sighted people could see the latter even in the unpolluted and clear desert skies of medieval Arabia, where this title was coined. We must remember that eyeglasses had not yet been invented! If you cannot see Alcor just beyond Mizar's glow, even on a very excellent night, then it is probably time to consider a visit to the eye doctor.

Selected Naked-eye Double Stars

object A, object B	*magnitude A*	*magnitude B*	*separation (in arc-sec.)*
Mizar, Alcor	2.4	4.0	708
Alpha-1, -2 Capricorni	3.6	4.2	376
Alpha-1, -2 Librae	2.7	5.2	231
Epsilon-1, -2 Lyrae	5.0	5.1	210
Mars, Epsilon Geminorum[a]	1.2	3.0	153
Nu-1, -2 Draconis[b]	5.0	5.0	62

[a]Temporary double "star." Split during conjunction on April 8, 1976, by Steve Albers—who believes he could have split the pair when they were considerably closer if he had looked sooner.

[b]Mentioned by Patrick Moore as having been split with naked eye—but probably this is an error (especially considering the pair's dimness).

Besides trying to split double stars, there are other—in some cases, more easy, glamorous, and thrilling—tests of naked-eye acuity in the heavens. There is, for instance, trying to perceive with the unaided eye the smallest or narrowest features you can on the moon. Like most of the other tests, this one really does involve visual powers in addition to simple acuity. But it is a worthy quest in itself. To attempt seeing with the naked eye as much as possible of the cryptic markings that mystified our pre-telescopic ancestors, that is at least a good exercise of the powers of observation. It is also (and more importantly) a kind of tribute to the beauty of the moon and its preciousness. W. H. Pickering developed for naked-eye observers a scale of increasingly hard-to-see features on the moon. We present here a chart of the moon with the features numbered according to the scale. Pickering suggested that very good vision could detect No. 10, and exceptionally acute eyesight could reach No. 11. He thought that No. 12, the Riphaeus Mountains, might be too difficult for any human eyes without optical aid. When is the best time to try seeing these features (and others)

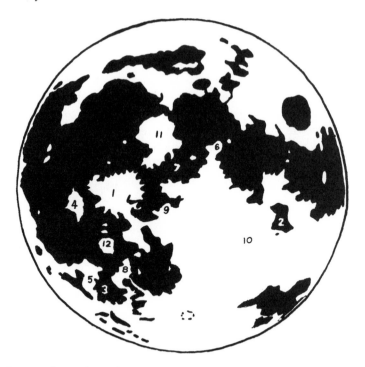

Lunar features for testing naked-eye acuity, numbered from 1 to 12 in order of increasing difficulty (after W. H. Pickering): 1—bright surroundings of Copernicus; 2—Mare Nectaris; 3—Mare Humorum; 4—bright surroundings of Kepler; 5—region of Gassendi; 6—Plinius region; 7—Mare Vaporum; 8—Lubiniezsky region; 9—Sinus Medii; 10—faint shading near Sacrobosco; 11—dark spot at foot of Appenines; 12—Riphaeus Mountains.

on the moon? During the period right around sunset (or, better yet, sunrise), when the moon is near first (or, at dawn, last) quarter. During the day, there is not enough contrast with the sky; at night, too much (the moon is a glare).

There are several legendarily difficult feats of astronomical observing that partly involve visual acuity. Among these are seeing with the naked eye several moons of Jupiter, the crescent of Venus, and the ring-elongated shape of Saturn. Each is a complex case about which much can be said. Here I can only stress my belief that seeing two of Jupiter's moons with the naked eye is quite possible with normal vision under the right conditions (I have done it), but that seeing the actual shape (as opposed to only elongatedness) of either the Venus crescent or Saturn shape without optical aid has not (to my knowledge) been documented, only claimed. Seeing the crescent of Venus is perhaps more likely to be possible. It certainly should be attempted when Venus is a few weeks from its "inferior conjunction" with the sun, and its crescent fifty minutes of arc and more tall. To avoid glare, try looking as soon after sunset as possible, or even before sunset, or (best of all) around midday! You may never succeed in this difficult quest, but far, far easier is locating Venus as a point of light in the noon sky. The trick is in knowing exactly where to look in all the immensity of bright blue sky. Even if you have a fair idea of where Venus should be in relation to the sun, binoculars or a small telescope (with very wide field of view) can be very helpful in locating the planet. Even better is using the daytime moon as a guide when it happens to be close to Venus. Remember to keep your eyes focused far when looking with the naked eye! If you follow these instructions, especially on a clear day when Venus is very near its greatest brilliancy, you will succeed in seeing this marvel the properly used eye can bring us: the sight of our sister planet together in the sky with the sun it orbits!

The matter of color perception in astronomy deserves special attention. Because the cones are responsible for perceiving color and do not function well at low light levels, the hues of most stars appear even more subtle than they might otherwise be. What is the faintest star in which you can perceive any hint of color with the naked eye? Most students of this question would say second or maybe third magnitude. Answer the question for yourself. A good way to practice is to rate colors of the brightest stars, and the best time is between autumn and spring, when the especially brilliant and especially colorful stars of the winter constellations are conveniently visible. Most people immediately notice a slight ruddiness or at least orange-yellow to the star in Orion's shoulder called Betelgeuse. The other first-magnitude star in Orion, Rigel, has less obviously some blue in its whiteness. What other colors can you glimpse? Try rating them carefully and skeptically using the accompanying scale. When comparing the colors of different stars, beware of the fact that greater brightness makes a star's color more prominent. Although a scale like this is not truly quantitative, it does aid in training your eyes to perceive and appreciate these lovely, elusive hues.

Your observations of star colors should be made on a night of good "seeing," when the stars are not twinkling too badly, or else you may be in for a surprise: an unsteady atmosphere produces rapid and various color changes in stars! This can be seen beautifully with the naked eye in Sirius, the brightest star, especially when the star is low or there is considerable turbulence. Rays of all colors (components of the actually almost white light) dart from the heart of Sirius at such times.

A telescope can reveal these atmosphere-caused colors in much fainter

A Scale for Rating Star Colors

 0. Very blue.
 1. Blue.
 2. Blue-white.
 3. White.
 4. Yellow-white.
 5. Light yellow.
 6. Deep yellow.
 7. Light orange.
 8. Deep orange.
 9. Orange-red.
10. Red.

Use decimals between the whole number ratings if warranted.

stars, but it can also reveal their permanent, natural colors. Authorities have often scoffed at claims that these colors are sometimes prominent. They point to the fact that in many double stars the seemingly strong hues are principally a result of contrast between the two different colors. Even if this is true, and the apparent colors are therefore misleading, what the eye perceives is nevertheless beautiful and interesting. Of course, color descriptions of all kinds suffer from imprecision, which results in part from our color vocabulary. Whether it be from language, observational habits, or even possible slight physiological differences, people describe and (in at least one way) actually perceive colors differently. Despite all this, the fact remains that quite a few true star colors can be perceived by the eye and their identity roughly agreed on. Even the scoffers acknowledge that very cool stars show fairly distinct color (ranging into yellow-orange and toward orange-red). And what about the reality of some double star colors? Antares, the "heart" and brightest star of Scorpius, is a slightly ruddy star which under considerable magnification and good "seeing" shows a much dimmer companion star that appears startlingly green. Is this merely a contrast effect, the companion's true color being a very pale hue? On at least one occasion when the moon was still in front of the ruddy main star and the companion was visible by itself for a few seconds, an experienced observer reported that it still appeared distinctly green. In the latter part of the 1980s there is a series of these lunar occultations of Antares for observers around the world, so perhaps this question of the color of the companion can be better answered.

The full range of colors we can see is said to be from violet to red—the colors we report when our eyes respond to wavelengths of light ranging from just under 400 nanometers (extreme violet) to well over 700 nanometers (extreme red). Some sources say that the extremes are 380 and 760 nanometers. Beyond the violet is the ultraviolet; beyond the red is the infrared. Bees and horseshoe crabs can see into the ultraviolet and therefore detect the sun's exact position even through moderately heavy clouds; snakes can sense into the infrared and thus form a rough image of their intended victim's heat in the darkest night. But can some people see a little beyond the typical human limits, into the infrared or ultraviolet?

Apparently there is significant variation from person to person as to how far down either end of the spectrum they can see. First, consider red. The human eye is far more sensitive to red light than most films are. Red stars show up on photographs relatively fainter than blue ones of the same visual magnitude. Perhaps some people are capable of seeing extreme red better than others to the extent that stars like Betelgeuse look ruddier. Independent of this question (yet not unrelated) is the interesting *Purkinje effect.* The nineteenth-century Bohemian physiologist Johannes Purkinje noted that the relative brightness of blues and reds depended on the illumination level, with blue becoming progressively more prominent relative to red at lower levels of illumination. The cause of the Purkinje effect is that the cones have their greatest response at about 560 nanometers, while for the rods it is about 510—the latter wavelength being markedly bluer.

There is certainly more clear evidence from astronomical observation that some people can see farther down the violet end of the spectrum than others. Young people can generally see more extreme violet than older people because the eye's lens not only becomes less transparent with age but also becomes especially less transparent to shorter wavelengths (the violet end). One observational effect of this is that young people tend to see many of the celestial objects called *planetary nebulae* as blue, whereas older people see them as green. These planetary nebulae are so called because some resemble the blue-green disks of Uranus and Neptune in telescopes. What they actually are is the ejected material of a dying star, and the remaining star is even more predominantly than the nebula a producer of very violet and ultraviolet light. A strange epilogue to the increasing loss of violet perception with age is provided for some people who have a cataract operation and have a lens replaced. The expert observer Walter Scott Houston had such an operation on one eye and afterward has been able to see the central star of planetary nebulae far better with that eye than with the other!

There does seem to be astronomical evidence, however, that violet-end perception differs also from individual to individual, independently of age.

My friend Steve Albers, an excellent observer, found that he was able with a little practice to see the extreme violet emissions at about 390 nanometers in some displays of aurora (Northern Lights). His photos confirmed his observations. I write "extreme violet," but aurora experts Aden and Marjorie Meinel call this emission ultra-violet and note that "To the eye this purple light is invisible but not to the camera." So Steve was seeing light and color that are commonly regarded as beyond the capability of human vision to glimpse! And perhaps it really is for many people, especially older ones. Yet I wonder if perhaps many or even most people are capable of exceeding the conservative limit of 400 nanometers. The aurora provides an almost ideal opportunity to test whether your limit extends below 400 nanometers. The easiest of this 390-nanometer emission to see is that at the uppermost part of some auroral displays, which is high enough to be in sunlight and thus fluoresce with added energy. But is there anywhere else in nature so fine a source of light specifically around this wavelength?

In the aurora, the 390-nanometer emission is from ionized molecular nitrogen high in Earth's atmosphere. But Albers and I have wondered if an extremely violet emission may account for an observational mystery about the most glorious "deep-sky object" (object beyond the solar system) for telescopes: the Great Orion Nebula. Many authorities have stated firmly that only a greenish glow from the two strong emissions of doubly ionized oxygen (at about 500 and 510 nanometers) is detected visually in this nebula even by observers using the world's largest telescopes. Yet many an amateur astronomer with a thirteen-inch (or quite a bit smaller) telescope has detected red and/or violet—colors that do show up on photographs of the nebula. Albers and I both see violet more often (red would presumably be the emission from hydrogen). Perhaps the reason why some admittedly expert observers have not glimpsed the colors is their poor violet-end sensitivity. The violet areas seem bright enough that they ought to show prominent color in a thirteen-inch telescope, yet Steve and I find the color strangely inconspicuous—again consistent with the idea that the difficulty is not too low an illumination for easy color perception but rather too violet an emission for easy color perception.

If you can glimpse a hint of violet in the edges of the Great Orion Nebula or the sunlit tops of some displays of the Northern Lights, perhaps you are really seeing a color beyond ordinary violet, unglimpsed elsewhere in nature by anyone—name that color what you like! Your attempts to see this color should be made as soon as possible, for you may lose the ability to see it as you age—just as we also tend to lose the ability to hear short wavelengths (high frequencies) of sound as the decades pass. The clear

message in any case—whether you study the colors of stars, nebulae, meteors, aurorae, twilights, or rainbows—is to be aware of hues in the heavens. There is not only great beauty but also seldom-appreciated knowledge to be gathered from them.

I cannot conclude this chapter without mentioning a few more of the human eye's abilities that are especially useful in astronomical observation.

I have already noted the eye's ability to adjust to the tremendous brightness variation of day to night by virtue of its two-fold system. But the eye can see over a great range in any one brightness environment, too. In order to show the easily visible rings of cloud-caused corona color around the sun or moon on a photograph, one must enormously overexpose the celestial objects themselves. Likewise, only composite photos can reveal the outermost glow of the sun's true corona (outer atmosphere) during a solar eclipse without overexposing the vastly brighter inner corona (and hiding the beautiful solar "prominences")—but the eye does this quite ably. On the other hand, the eye is also good at distinguishing very slight differences in brightness. In this ability it falls far short of today's ultra-sensitive photoelectric devices, but veteran observers of *variable stars* (stars that change their brightness) can distinguish between stars whose brightness is as little as 0.1 magnitude different. In a country sky, then, the naked eye can note as many as 45 different brightness levels between magnitude 2.0 and 6.5— plus a rich scattering of brighter stars and many various degrees of brightness in the gorgeous, intricate glow of the Milky Way band.

Mention of the Milky Way leads us to consider the extremely wide field of human eyesight. As previously noted, our vision is not very sharp beyond a small area at the center of our field of view. But the sharpness is great enough and, more significantly, the eye's natural scanning motion fast enough to give us fine sights of entire giant constellations (or groups of them), long stretches of the Milky Way, much of the activity of a meteor shower, huge twilight glows, and so on—all far better views of these particular things than can be provided by the widest-field telescope or binoculars. A few camera lenses provide a wider (in fact, even an all-sky) panorama than our eyes can, but bright planets and stars can be glimpsed even out in the peripheral parts of our vision at almost opposite corners of the sky. The author John McPhee had then–college basketball star (now U.S. Senator) Bill Bradley tested by an optometrist to see whether his field of view was wider than the average person's (if so, it would be an advantage for a basketball player). According to the doctor, the logical (rather than experimentally established) norms for eyes looking directly straight ahead are in the horizontal 180° and in the vertical 112° (47° up and 65° down).

Most people the optometrist had tested actually had *less wide* fields of view, but Bradley turned out to have 195° horizontally and 140° vertically—70° down but also an amazing 70° up (about 23° better than the expected logical norm!). The doctor doubted whether practice at trying to see objects with peripheral vision could actually improve one's effective field of view, but Bradley said that he had indeed practiced this while growing up. An interesting and seldom-tried exercise for astronomical observers is to see how wide a panorama of constellations they can behold without moving the eyes, and how far apart they can see bright planets and stars in one view. The full width of the human field of vision cannot be exploited for even fairly bright celestial objects, of course. But looking at some of the spans of many or all planets in the sky at once in the early 1980s proved to many of us that the effective field is nonetheless very great.

One kind of object with many characteristics best revealed (or only revealed) by the naked eye is comets. When a comet's head is bright enough for the naked eye, the size and magnitude of that elusive cloud is almost always greatest for the viewer without optical aid. Comet tails are also sometimes best with the naked eye—if they are long, in fact, they cannot even be fit entirely in the field of a pair of binoculars. Most of Halley's great dust tail was never positioned so as to have much surface brightness at this most recent return. But the naked eyes of a few Halley observers in late April and early May 1986 were able to trace out an immense length of the tail which may have been longer than anything the cameras and electronic light detectors achieved.

The human eye is difficult to surpass as a perceiver of dim extended glows in the heavens. Any optical instrument's ability to reveal such glows depends partly on what is called (following photographic parlance) its "speed." The speed is measured by the instrument's focal ratio (or "f-ratio"), and that is the ratio of its primary lens's or mirror's focal length (the distance from it at which it brings an image to focus) to the lens's or mirror's diameter. Thus a telescope with a focal length of 48 inches and a mirror diameter of 6 inches possesses an f-ratio of 48 divided by 6, which is 8 (usually written f/8). The lower the number of the f-ratio, the "faster" the telescope and (generally) the greater a surface brightness of an image it can present in its range of usable magnifications with eyepieces. The lowest f-ratio and therefore "rich-field" telescopes widely available on the market are f/4. But the human eye's f-ratio is far lower! It does not have these telescopes' light-gathering power or magnification, but it does perform some of the same magic that observers associate with those instruments— and more of its own. Rich-field telescopes can often show nebulosity (cloudy, glowing dust and gas) around the lovely Pleiades star cluster, but in recent years Walter Scott Houston has gotten many reports of naked-eye sightings of this glow. Sometimes an observer must have difficulty ascertaining what is really nebulosity in even an excellent naked-eye view of the cluster because of its close bunching and multiplex twinkling. But apparently these contemporary observers have proven that there really is an added significance to Alfred Tennyson's unforgettable lines from "Locksley Hall":

> Many a night I saw the Pleiads, rising through the mellow shade,
> Glitter like a swarm of fireflies tangled in a silver braid.

Who would have ever guessed that the powers of human vision were so great? Our unaided eye can see five times as many stars as we thought, a color we did not know existed, moons of other planets, and a planet together with its parent star in the daytime sky. Our unaided eye can see gleams 10^{14} times more faint than the brightest it can stand, roughly 5 million different shades of color, detail as fine as a pencil a quarter-mile away or a person nine miles away or a minor spur of mountains a quarter-million miles away or the gap between two suns in the same solar system hundreds of light-years away.

I wrote earlier that the powers of the eye are far-ranging. I meant it in several ways, including the most literal. How far can the eye see? More than 2 million light-years distant without optical aïd, and a significant fraction of the way to the light-speed-limited "edge" of the expanding universe with telescopes. On a clear night we cannot quite see "forever," but when our unaided eyes glimpse two dim patches of radiance overhead in America, Europe, Russia, or Japan on autumn evenings, these eyes are time machines showing us light that left the Andromeda and Triangulum galaxies more than 2 million years ago—the time when human creatures on this world were probably first opening their eyes. So these humble gleams (really the combined mights of maybe a trillion stars each) connect the first truly human eye to the most recent in an unbroken line which shines across even *that* fearfully inconceivable—but triumphantly vision-bridgeable—abyss of intergalactic space.

In the categories of light-gathering power and visual acuity, we might envy some creatures: hawks with their binocular-power eyes, snakes with their heat-form images, owls with their million-star skies, a basketball-playing senator literally more wide-eyed than the rest of us. Some owls can see far more sharply than we can even by day, and some other birds have two foveas or better-than-Bradleyan wide fields of view. But would we really trade our *combination* of visual powers for those of any other creature? Few birds have supremely excellent vision for both day and night, and apparently almost all birds have very poor perception of blue and violet.

When all has been considered, even the spectacular competition of our fellow creatures' vision may not stand up to human vision—at least not to its title of being the best all-around instrument for gathering information about this planet and universe. In light of the beauty, majesty, and interest of our world and its cosmos, there could be no higher title for a mere thing, no better physical gift for us.

And nothing more exciting and potentially joyous to say about a thing than . . . *I see it!*

11

The Next Supernova

THERE IS A FAIR CHANCE that many people alive today will get to see the next supernova in our galaxy. That star of stars will illuminate not only the heavens but also our understanding of stars and stellar evolution as no other natural event could. That one mighty flash of star, and its glimmerings after, will teach us vastly more about all stars.

But even the immense harvest of knowledge we could gather from supernova-light, from the greatest possible explosion of a star's contents to the wide universe, might not by itself be the greatest benefit. For a supernova could strike straight through to one thing at the heart of all great intellectual and spiritual adventures: let us call it, in this context, wonder.

Recent historical research has revealed eyewitness accounts of brighter and more wondrous supernovae than were ever known before. This essay is about what a bright galactic supernova will look like and how the unleashing of that splendor could inspire us not only to greater insights and appreciation of the starry universe we live in but maybe even also of one another.

OVERDUE FOR SPLENDOR

Many of us who are interested in astronomy have read about the classic supernovae of 1572 (Tycho's Star) and 1604 (Kepler's Star), as well as the 1054 supernova, which gave birth to the Crab Nebula. To these famous objects can now be added about seven more supernovae that were unverified or even unknown to modern astronomy just a few years ago. Much of the data in this essay are derived from the work of David H. Clark and F. Richard Stephenson presented in their book *The Historical Supernovae* (1977), a delightful mixture of clearly explained astrophysics, Chinese history, and supernova eyewitness accounts that reads like an exciting detective story (because it is!). On the basis of excellent work like this, and some simple (though perhaps previously unthought of!) considerations of our own, we can get a vivid picture of what overwhelming visual beauty could await us at the coming of the next supernova.

The "new stars" of 1572 and 1604 peaked at brightnesses similar to those of Venus and Jupiter. Most people who know anything about this probably still imagine that a supernova in our galaxy would be much like having another Venus or Jupiter in our skies for a while. Although "another Venus" is certainly a marvelous prospect, it falls short of conveying many aspects of such a star's splendor. More important, a galactic supernova at maximum could easily be an object of much greater brightness.

Of course, some historical supernovae were much fainter than those of Tycho and Kepler. Large amounts of interstellar dust in some regions of our galaxy have masked their true brilliance. If we could see it clearly, even a supernova on the extreme opposite end of our galaxy would shine at a magnitude of +1 or +2 (at least as bright as a Big Dipper star). A few of these dust-dimmed supernovae have almost certainly been missed, especially before modern times, since even the diligent and dependable Chinese observers would not have seen a supernova peaking much fainter than +1.5 magnitude. Today's amateur astronomers, many of whom are regular watchers of "variable stars," have a much better chance of noticing such an object and could provide an absolutely invaluable service to science by doing so. (This is yet another strong reason why the nonprofessional like you or me should be out in that back yard watching the heavens.)

Of course, if the next supernova is one of these dust-dimmed cases, it will probably not much impress the general public. But backyard astronomers would still view it with considerable awe because of their appreciation of its nature. They would know, even if it appeared fainter than the North Star or a fairly close *nova* (lesser kind of star explosion), that this single star did behind the light-years of dust actually rival the combined light of a hundred billion suns! This seemingly gentle flicker in our sky would have utterly destroyed whatever planets might once have orbited it. Yet we would know it to be simultaneously a star gone to seed—a star spectacularly sowing space with heavy elements and the promise of new stars, worlds, life, and eyes.

But the historical record reveals that we are probably a century or two overdue for a supernova. And not just for any supernova, but for a brilliant one. The lack of dazzling supernovae between A.D. 185 and A.D. 1006 might be the result of sparse records from troubled times (even the records we have from the Chinese may have large omissions). It may also reflect the

Possible and Certain
Historical Supernovae in Our Galaxy

Year of Appearance	Months of Naked-eye Vis.	Max. Brightness	Location (1950 coord)		Comments
			R.A.	Declination	
185[a]	20	−8	—	—	near Alpha and Beta Centauri
386	3	+1.5	—	—	in Sagittarius
393	8	0	—	—	in Scorpius
1006[a]	24	−9.5	15h 00m	−42°	in Lupus
1054[a]	22	−4.5	5h 32m	+22°	produced Crab Nebula
1181	6	0	2h 02m	+65°	in Cassiopeia
1408	4	−1.5	20h 00m	+35°	probably Cygnus X-1 radio source
1572[a]	16	−4	0h 23m	+64°	Tycho's Star
1604[a]	12	−3	17h 27m	−22°	Kepler's Star
1680	brief	+4	23h 21m	+59°	Cassiopeia A radio source

[a]The most certain supernovae.

paucity of modern astronomers who can, or have cared to, research and translate historical documents. The records of the 1408 supernova were identified as late as 1979 and that one was apparently much brighter in the summer constellations than summer's brightest star, Vega, for at least a few weeks.

Furthermore, the study of supernovae in other galaxies also suggests that the present dearth in ours may be unusual—just a statistical fluke. Being "due" or "overdue" for this splendor does not mean that a galactic supernova (supernova in *our* galaxy) is guaranteed to appear in the next twenty, fifty, or even one hundred years. We are dealing only with averages. It is also true that our data are still quite limited. And yet we cannot help but be encouraged by the figures and the apparent odds. We do seem to have a fair chance of seeing a bright supernova in our lifetimes.

How bright will that supernova be? The historical record compels us to give free rein to our imaginations, for at least twice in the Christian era there have been "temporary stars" which rivaled or surpassed the brightness of a thick crescent moon.

A STAR AS BRIGHT AS A
HALF MOON

Just forty-eight years before Earth saw the Venus-bright (or slightly brighter) creation of the Crab Nebula, a star appeared that put that supernova of 1054 to shame. In the spring of 1006, in the constellation of Lupus the Wolf, a supernova reached a maximum apparent magnitude of about −9.5—roughly the brightness of a half moon! This magnitude estimate is based on eyewitness descriptions, especially those of its illumination of the landscape.

It was first spotted when it "only" rivaled Mars, which was then near opposition at about magnitude −2. In the next few weeks, the supernova burned steadily brighter and brighter until it outshone Venus, then brighter still so that it illuminated the sky all about it. Finally it became so searingly brilliant that its radiance had influence over literally the entire sky.

We do not know whether the Lupus supernova was of Type 1 or Type 2. Type 1 is the rarer and intrinsically brighter of these two most important varieties of supernova. The types have somewhat different light curves (lines on the diagram plotting brightness over the course of time). If the Lupus object was the more common Type 2 supernova, the star should have remained brighter than −8.5 magnitude for about three weeks, and brighter than −7.5 for more than three months. In all, it would have remained distinctly more brilliant than Venus for around five months! Such a supernova would have been—and apparently was—visible to the naked eye for more than two years.

These figures are impressive, but they take on full force only when we consider the individual observational effects produced by such brightness.

The effects would not be merely greater than those caused by Venus or Jupiter. They would be different—phenomena that no bright planet or ordinary star could ever imitate. They would be unfamiliar to the most experienced observers and have the wonder of the unique and new.

So let us imagine now that we are not just a little lucky but instead extremely fortunate. Let us imagine that the next supernova occurs in the relatively near future and shines roughly as bright as the star of 1006. For convenience, we will call this next supernova Lupus 2. With the aid of the eyewitnesses who actually saw the star in 1006, let us picture what the star in our future would look like.

For the brief period of maximum brilliance, Lupus 2 would illuminate the landscape to the same degree as a first-quarter moon. At night we could see distant objects by its light alone. We could read the print of a magazine or newspaper or book by its light, and we could see some dim colors in the landscape.

Lupus 2 would cause prominent shadows for several months. Under the right conditions Venus casts shadows, so a −4 magnitude supernova in a dark midnight sky (where Venus can never shine) probably would too. Lupus 2, more than one hundred times brighter, certainly would.

Lupus 2 would literally light up the whole sky. A half moon affects the whole sky with its radiance and hampers our seeing even rather bright stars in its general region of the heavens. Lupus 2 would cause similar problems for observers of other celestial objects, but for many, many weeks without respite (unlike the moon, which dwindles to a crescent and invisible new moon and moves on soon to permit undisturbed viewing of particular constellations and other objects). In this case, however, neither backyard astronomers nor professional astronomers would be likely to complain!

To truly appreciate these comparisons, go out on the next night the moon is at first quarter. But remember, what we have so far considered about Lupus 2 is comparatively tame. Only when we consider the *differences* between a supernova and the moon or bright planets can we understand what staggering beauty will be unleashed upon us.

Like any star, the supernova will appear as a point of light. The light of the moon, by comparison, is spread quite dully over a rather large area. But the supernova's brilliance will be intensely concentrated. Since a light source of magnitude −19 has been shown to be painful to look at in laboratory experiments, Lupus 2 could be an excruciating beauty in large telescopes.

For the same reason (concentration of the light), the Lupus 2 supernova would be far easier to see in the daytime than the moon. On a clear day when Venus shines as bright as magnitude −4, it is plainly visible to the naked eye—but only if you know exactly where to look in all that big, bright sky. A much brighter supernova would probably be spectacular in the daytime sky, and Lupus 2 would be hard to miss—it would be more than a mere candle in sunlight. In the daytime, Lupus 2 ought to seem roughly as bright as Sirius—the brightest star—does in the night! Of course, day and night observing are not analogous in many respects—but consider that the great supernova of A.D. 185 was prominent enough that it was discovered from Lo-yang in China when it was *extremely low in the daytime sky*. Perhaps the idea of Lupus 2 as a "Sirius of the daytime" is not an exaggeration.

Any extremely bright supernova would create cloud-coronae. Coronae are not halos, vast circles sometimes seen around the sun and moon; they are much smaller colored areas right about the sun or moon (they are seen most commonly in clouds passing in front of a bright moon). Lupus 2, or even a somewhat less brilliant supernova, would surely produce coronae. These would probably be quite unlike the moon's, and lovelier, because a stellar source would cause smaller, sharper bands of color. I saw a corona something like this at the end of a total solar eclipse, and those few seconds may have been the most spectacular I have experienced in a lifetime of sky watching.

Then there is an effect caused by the structure of the human eye itself. While the concentration of so much light within a perfect point would be beautiful, defects in the human cornea and pupil-edges produce also beautiful "rays" or "spikes" extending from sources like stars. The brighter an object, the longer its rays appear. How big would Lupus 2 appear compared with Venus and Jupiter? How far would its rays appear to extend as seen by the naked eye? The size of any given source's ray-spread must vary a bit from person to person (and eye to eye), also according to other factors like the level of general illumination. Nevertheless, we have estimates by skilled observers from before the invention of the telescope which tally well with those of today's observers. And in connection with the Lupus supernova of 1006 we have a fortunate circumstance: a statement on the object's apparent size by an observer thought to be highly reliable.

That observer, Ali Ibn Ridwan, said the supernova appeared "two and one half to three times as large as Venus," meaning, of course, the width of its spread of rays. Since various observers (including me) have estimated the apparent ray-spread of Venus as roughly 3 arc-minutes, we can conclude that the 1006 supernova was as large as 7.5 to 9 minutes of arc across. That is almost a third as wide as the full moon, a figure which may not sound tremendously impressive until you actually go out and look at the full moon and picture the star. What does seem—and is—tremendously impressive is to picture three mighty Venuses as close as possible side by side and know that this span was that of the Lupus supernova, but in all directions, and with a radiance the equal of one hundred Venuses.

THE STAR-FOUNTAIN

In this exploration of what Lupus 2 would be like, we have neglected one property which supernovae share with all stars but not with bright planets and the moon. Any star seen through the Earth's atmosphere *twinkles*.

There has never been a context in which this wonderful fact has been so important and exciting.

With the magnitude and a half step from bright Vega to brighter Sirius, we see a wonderful increase in the noticeability of twinkling and color-variations. Some people have never recognized the color changes in the twinkling of any star but Sirius because the phenomenon is so much more prominent in this by-far-brightest of stars. If you have never observed the phenomenon yourself, go out early on a clear December evening and view Sirius when it is low in the sky (the twinkling and color-changes are caused by turbulence in our atmosphere and are always strongest when the object has the longest pathway of air to shine through while low in the sky).

Now imagine a star far brighter compared with Sirius than Sirius is to Vega or to Rigel, the bright blue star in Orion's west knee. Such a star would still only rival Venus in brightness, but the planets hardly twinkle or change color at all. The awesome twinkling of a -4 magnitude supernova alone would make it very different from and very much more than just "another Venus or Jupiter."

How then can we imagine what the twinkling of a moon-bright, multi-Venus-size Lupus 2 would be like?

Try to picture supernova-rise. Only an extremely murky night could quell the radiance of such a star even at the horizon. It would be an endlessly dancing and varying mass of fire, shooting spears of intense color every way, a fountain of starlight heaved in one many-waved leap over the gap of several thousand light-years. Unlike the dull-bodied moon, it would often be spectacularly visible the *second* it came above the horizon—one instant deep darkness, the next its many-colored light bursting over all the sky and landscape in a flash that would make the heart leap with beautiful dread and wonder. As it rose, Lupus 2 would quickly gain still more power and flood every corner of the night with pulsing illumination. On ground, tossed flocks of racing "shadow bands"—those strange strips of shade seen for a brief time just before or after some total eclipses of the sun—would animate the landscape. At latitudes where the star was always low, or on nights of unsteady atmosphere, the shadow bands might rove eerily through forest and town from dusk until dawn.

Its mad scintillation, its gliding shadow bands, its changing colors and changing cloud-coronae of different colors would be at times frighteningly strange and awesome. A bright supernova would in some ways be like a total eclipse of the sun, in others like the most striking of conjunctions—but the drama would last not a few minutes or nights but for many months.

And what about combining the drama of a changing conjunction or eclipse or occultation with that of the supernova? Supernovae are most likely to appear fairly near the band of the Milky Way in our sky, a fact itself promising of further beauty. But what are the chances of one being near enough to the ecliptic (the middle line of the zodiac) to have conjunctions with the planets or the moon?

No less than three of the seven best-verified supernovae lay in the zodiac, and at least two of those were involved in close conjunctions. The Crab Nebula supernova appeared next to a crescent moon when it first burst into great brilliance, and Kepler's Star appeared not many degrees from a fine gathering of Mars, Jupiter, and Saturn—a coincidence which led to the supernova's discovery when its brightness was still on the rise. Today, an early observation would be of tremendous value, and it is the backyard astronomer who is most likely to make it. Will the next such discovery be of a supernova in conjunction with planets? The heavens are so rich that a conjunction with some kind of interesting object seems highly probable.

Perhaps the next supernova will appear close to the sun. If it were Lupus 2, the supernova would more than hold its own until the sun moved off along the zodiac, leaving the star in a night sky. But think of the glory of Lupus 2's rising just minutes before dawn—a greatest of all Morning Stars heralding the sun's coming. Although the odds are very much against it, suppose we got to see the edge of a crescent moon passing right in front of Lupus 2, instantly plunging the night into a deeper darkness? Even if Lupus 2 were only in the same general vicinity of the sky as a crescent moon, we could be thrilled by the experience of a unique sight: the lustrous moon immensely outshone by a single, glittering, intense star.

If Lupus 2 grew to brilliance beside only an ordinary star, the conjunction could still be lovely. The A.D. 185 supernova shone just a few degrees from Alpha and Beta Centauri. Those bright stars must have dwindled into seeming tiny children of the supernova when it ignited up to magnitude -8. But as the exploding sun faded, there must have been stage after stage of different, striking appearances of the three brilliant objects gathered here. (Alpha and Beta Centauri are the best and closest pairing of two very bright stars in all the heavens, much better than northerners' Castor and Pollux in Gemini—what an amazing coincidence that a great supernova should occur so near to this pairing, making it a still more marvelous *trio!*)

For those of us who love the constellations and their legends, a supernova making majestic alteration to Leo the Lion or Aquila the Eagle or strong Hercules in the sky would be a rich thrill in itself.

The possibilities are almost endless. I think of Lupus 2 remaining visible even to within a few degrees of the sun. That would be much like one of the awesome "sun-grazer" comets, but they can never shine with such brilliance, even for a few days, when far from the sun in the sky. Lupus 2 could. It would simulate the experience of being on some planet of another solar system where two suns shine brightly in the sky at once. And I keep

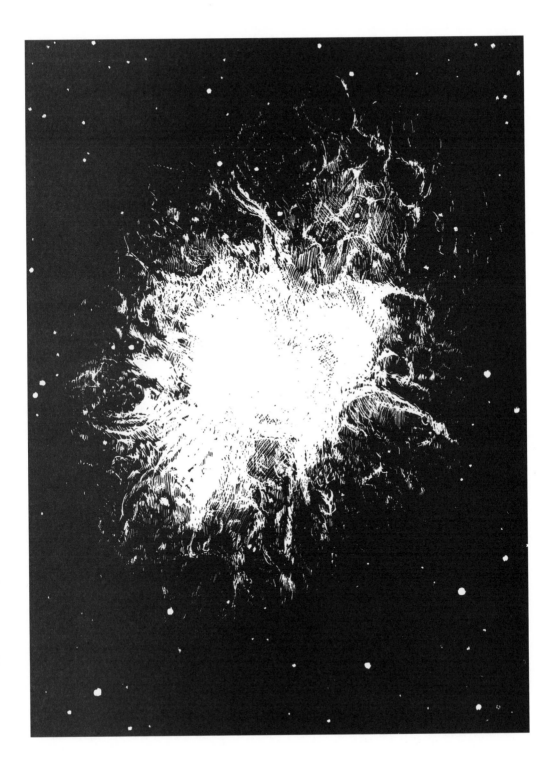

thinking back to the beauty of a star which could shine with great power and colors even right in the midst of the lovely hues of dawn it was rising in just before the sun.

CONCLUSION

If our next supernova is strongly dimmed by interstellar dust, it will be visually exciting only because so intensely interesting and important to our knowledge of all stars. If it is in the medium range of brightness for such objects, it will for several months be not just the most fascinating but also the loveliest object in all the heavens. But what if the next supernova is one of the brighter ones? Then it could be the most staggeringly beautiful heavenly object we can imagine.

According to Clark and Stephenson, a supernova of apparent magnitude −14 (much brighter than the full moon!) might appear roughly once in every 200,000 years. That is not often, but presumably human beings have seen such an object—maybe a number of times. There is an argument that a supernova far brighter than the one in Lupus may have occurred as recently as 4000 B.C. (but probably sometime in the five or six thousand years before that) and created the Vela pulsar. We do not yet know enough about these possible ancient supernovae. We do know, however, that in the past 2,000 years there have been supernovae of half-moon-rivaling brightness twice—and that we may be due for one again quite soon. What kind of impact would a Lupus 2 now have on the people of the world as a whole?

No one can answer that question adequately. We might hope that we would see little of superstition and commercialism in regard to this splendid object (if it were a south circumpolar object, the southward exodus to see it would probably be a thousand times greater than that for Halley's Comet in 1986—picture the travel posters declaring, "Come dine in Rio by supernova-light!"). We might also hope—perhaps not in vain—that this wild and stunning beauty would have a tremendous unifying and inspirational effect on the people of the world.

Few things are more sobering than how long it took for us to rediscover that such events as the Lupus supernova even took place. It is true that there was in 1006 nothing at all like the instant communication and world-wide consciousness of current events that we have today. But the forces which destroyed or eroded away the records and memories are the same that threaten our world now: greed and selfishness and aggression, short-sightedness that sees not the needs of neighbors next door or children tomorrow. What a shame that the splendor of the great stars of the past was so quickly forgotten. Perhaps the next supernova will also shine ultimately unheeded, in bitter beauty over a desperate or deserted world.

But perhaps we have reason to believe otherwise. Although the threat to us has never been so great, neither has the awareness of it and the voice and action of men and women of good will against it. We have come a long way since 1604—the time of the last supernova we saw in our galaxy—and even farther since the last Lupus supernova. The next supernova beckons to us from the future—perhaps tonight!—as a symbol of what we should strive for. If we keep its image in our minds, we will improve the perceivers (be it we or our children or our further descendants) who greet it when it finally does appear. When it comes, it will overflow the night with radiance, and its fire may kindle more than our eyesight: it may kindle our hearts. This time, the "guest star" could be seen, photographed, painted, shared, studied, enjoyed—and never forgotten. When the next supernova appears in the heavens, we would finally be ready.

12

The End of the Stars
(Not One Child in Ten)

ONE DAY I WOKE and found that almost half the county I had always lived in was gone—almost all of that beautiful half called the night. Ten miles from our small cities, forests still stretched almost uninterrupted, wildlife flourished, even eagles could occasionally be seen, the air and water were still fresh and pure. But even that far out, and much farther, a longer-reaching pollution extended. A glaring reminder of man's wastefulness and obtrusiveness, it had bruised much of the clear face of the night sky beyond recognition with livid blotches that would not go away, that were worsening and spreading one to the other like a cancer.

How could we let this happen to that most glorious and compelling and inspiring of nature's faces? If any sight in nature has always awed mankind, it has been that of the starry sky in full splendor. The very pinnacle of inspiration, the highest literal and figurative goal of humanity, had always been the stars. The most exalted vision that any single writer has maintained for the space of a long story is Dante's *Divine Comedy,* whose three books each end with the appeal to and naming of the stars. To Dante, starlessness was one of the attributes of the Inferno, of Hell. Meanwhile, Americans who love their country look at their flag and hope for "the Stars and Stripes forever"—while the real-life stars are fast disappearing over the land. People refer now to sports and entertainment celebrities of increasingly brief duration as "stars" while the age-old enduring stars of the heavens, fit objects for admiration, are being hidden out of sight and out of mind. Not just in my home county but everywhere in America, and the world, the destruction of the night sky is in full swing. In many places the disfigurement of that fair visage is essentially complete.

It is time for all of us to wake up, and not just to the damage but to answering action. Where you sit and read these words now, the stars may have already ended. Within the past decade or so, the stars have died or faded toward death for a majority of the people in the United States and Canada. Within the next decade or two, the dying will be complete in all but wilderness areas: the stars will exist as little more than a recollection of legendary beauty and inspiration—*unless we do something.*

What we have to do something about is "light pollution."

Light pollution is the killer of stars, meteors, comets and Milky Way, of all the thousands-of-years-old glory and significance of the constellations.

Light pollution is not all outdoor man-made lighting. It is the excessive and especially the misdirected outdoor lighting that does no one any good. We are not talking about a choice between either seeing the stars or having modern, technical civilization. Light pollution hurts both things, both aesthetically and practically. Besides its inestimable aesthetic or spiritual importance, "seeing the stars" is of great practical importance to the advancement of science (professional astronomers stress that observations from outer space cannot yet and may never meet all their research needs). As for modern civilization, the thought of a safely and attractively lit city with a sky full of stars above it is a beautiful (and, as we shall see, not an impossible) one; the practical problem of billions of dollars being lost to energy waste is, of course, one that hits home to all of us, whether we bother to look at the sky or not. We are all being robbed blind in not one but two ways: robbed of both the stars and our money. And in a sense, even money is not as practically important as energy in this world where nuclear power is difficult (some think impossible) to make safe, and international relations are perilously strained over control of remaining fossil fuel deposits. Is it an exaggeration to say that light pollution has a significant impact on even these issues of worldwide, gravest practical concern? Probably not, considering that light pollution may be the largest single waster of energy in the United States (and probably most other industrial countries).

So perhaps the pink sky glow of ever-brighter high-pressure sodium lights, soaking to sickening saturation the heavens for scores of miles around cities, really is as ominous as it looks. Yet it is these very same practical dangers—the lost money and energy and the world problems they exacerbate—that give us hope for action against light pollution. However business-minded our leaders and many of our fellow citizens may be (as if beauty and knowledge and nature were not part of man's most important "business"—that of life), they should be interested in learning about light pollution because of its economical and political significance. The key is getting the message out to them.

If you are already an active amateur astronomer, you are almost certainly all too well aware of light pollution's effects on your sky. But some of the most devoted and veteran observers are too cynical about the hope of ever checking the spread of light pollution. The last two of three major topics in this chapter are therefore the most important. I will discuss first just how severe, widespread, and rapidly worsening light pollution has become—in case there are any rural sky-gazers reading this who still suppose (as I think I long did) that the wave of glare will not reach their private heavens (or that ever better filters and larger telescopes will save their view). Second, I will present the three most important principles of good outdoor lighting and how they have been implemented in many places with strikingly successful reforms—to everyone's (not just astronomers') general satisfaction. Third, and as important as the second, I will tell the surprising story of my international Dark Skies for Comet Halley project and how individual citizens can not only publicize the need for outdoor lighting reforms but also help see that those reforms are implemented in their home areas.

HOW SEVERE IS
LIGHT POLLUTION?

A few years ago, an editorial in a major newspaper chided certain astronomers for a particular comment they had made about Halley's Comet. These silly astronomers, it seems, were saying that the 1985–86 return might be the last at which city lights would permit the comet to be visible from anywhere on Earth. Surely the glow from cities could not grow so strong that even rural observers would be prevented from seeing the comet in 2061, when it will be substantially brighter than it was this time? Little did the editorial writer realize that light pollution across all but the world's more remote wildernesses could be that bad in twenty years, let alone seventy-five. Few people, even amateur astronomers, appreciate the extent of the problem, and how fast it is spreading.

One way to help convey the reality is with photographs. On these pages you can see for yourself the brightness of city lights and the tremendous increases over even a matter of a few years. When a strong glow is noticed in one direction from your home or observing site and that glow begins a noticeable climb up the sky, the time left for any of your sky to be optimum is very limited. Even as I write this, yet another letter arrives here, this one from Richard Van Dyke of La Grange, Illinois, saying: "One year ago the night sky to the east (Chicago) was orange, and the rest was blue-black, with most stars visible. This year the sky is orange to the east, north and south, and the western sky is close to gray."

Perhaps most telling of all are satellite photos of entire countries. The famous "night face of North America" shot is impressive. But even this photographic image of the United States and southern Canada splotched and networked with glare—mostly waste—falls far short of conveying the extent of the problem. For one thing, the white splotches are only the areas absolutely inundated by light pollution. If you take each one and imagine it spread over a much larger area, you have a better picture of how much of America's skies are seriously degraded by light pollution.

How far from a city of given population must you be to avoid significant light pollution? The accompanying tables give figures based on Walker's Law, a formula which works fairly well for most cities, though it falls short of predicting the severity of light pollution in quite large cities where more than 1,000 lumens of illumination per person may be present. As you can

Continental United States at local midnight—USAF satellite mosaic taken in 1979.

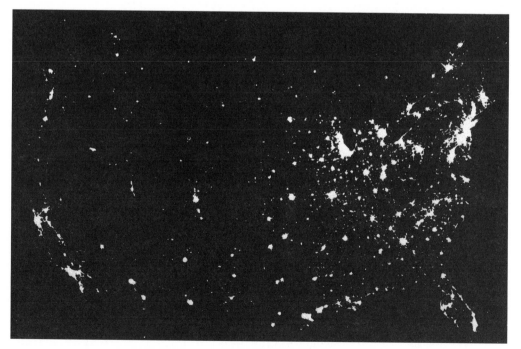

see, a city of a million people can have a serious effect on the night sky more than 60 miles away. Usually, though, it is the suburbs closer to you or your observing site which do the most damage, because closeness is relatively more important: a suburb producing only one-sixth as much light pollution as a big city will affect your sky just as badly if it is only one-half the distance away.

Light Pollution from Cities as a Function of Population and Distance

Distance at which significant observing degradation is beginning (sky glow 10 percent above natural background halfway up the sky in the direction of the city) can be matched with population of a city that would cause such degradation:

Distance (km/miles)	Population
10 km/6.2 miles	3,160
25 km/16.5 miles	31,250
50 km/31 miles	177,000
100 km/62 miles	1,000,000
200 km/124 miles	5,660,000

Effect of distance on the fall-off of sky glow (relative to "1" at 100 km):

Distance (km/miles)	Light Level
10 km/6.2 miles	316
20 km/12.4 miles	56
30 km/18.6 miles	20
40 km/24.8 miles	10
50 km/31 miles	6
60 km/37.2 miles	4
80 km/49.6 miles	2
100 km/62 miles	1

"The same source at half the distance has 6 times the effect." Data from Dr. David Crawford; calculated from formula of "Walker's Law."

● ● ●

A second fact to bear in mind about the "night face of North America" photograph is that it was taken way back in 1979. For several decades now, light pollution has increased at a much greater rate than population.

The 1960s seem to have marked a beginning to the real explosion of light pollution, for several reasons. Increased suburbanization was one. Generally lower electricity costs helped extend to outdoor lighting the untrue (but pervasive) philosophy that bigger (or brighter) is always better. Another important factor in the increase of the 1960s was the widespread introduction of the brighter-per-dollar mercury vapor lights. The mid-1970s saw trouble with Arab oil and greater concern about energy, but by the end of the decade conservation was slackening, and the usually brighter (because again more cost-effective) sodium lights were introduced. Light pollution has spread across the United States and the world like wildfire. In the 1980s the vast light-polluted areas around U.S. cities have extended so far that they have all begun to merge.

So we in the United States have reached a stage at which a nationwide qualitative change is occurring. Even if we consider the vast areas of sparse population in parts of the Great Plains and Rocky Mountains, most of the continental United States is now significantly light polluted. A large majority of Americans live under the dome of glare. The large national parks and forests of the western United States have been suggested as dark-sky areas where man-made lighting would be kept to a minimum and to which people could make pilgrimage to see the stars. But how many of us, how often, could take advantage of this opportunity? Furthermore, without strong efforts to reduce the light pollution from cities, not even the most remote wilderness areas can remain completely protected from the long reach of light pollution.

How long is it before almost all America's skies become the way those of large cities are now (but do not have to be)? In such cities the only heavenly objects readily visible are the moon and a few bright planets, when these are up. Even the brightest celestial bodies have little of their beauty left in the unnatural, livid, and ugly city skyglow. In these areas the separation of human from not just Earthly nature but from the greater universe and its higher beauties is virtually complete. The sky by night or day is becoming choked, lost, and forgotten. As Guy Ottewell has said, life is becoming for most people a succession of the insides of cars and buildings.

Here is another way to think of it. *If light pollution continues to increase at anything like its present rate, not one child in ten being born in the United*

States today will ever really see a star. What kind of people will be those who have never had this experience? I hope we never find out.

PRINCIPLES OF GOOD OUTDOOR LIGHTING AND LIGHTING REFORMS

In recent years, a small number of American cities, some very large, have passed reforms in outdoor lighting practices. No city has yet done anywhere near as much as it could do to reduce waste lighting. But the laws that have been put into effect have made a significant difference in skyglow (especially from what could have been expected without reform) and have met with general approval by almost all elements of society.

There are three major principles of good outdoor lighting upon which reforms can be based. Outdoor lighting should be (1) only *where* needed, (2) only *when* needed, (3) only the *kind* needed. Leading examples of reforms that implement these principles are (1) full-cutoff shielding, (2) ad lighting curfews, and (3) low-pressure sodium lighting.

Directing light only *where* it is needed must be the most important practice for economics—and for astronomy, since man-made lighting is seldom needed in the sky! Nor is it needed shining directly in a driver's eyes on the streets. Full-cutoff shielding directs all light down to where it belongs rather than partly up or to the sides. Such shielding can reduce energy needs, costs, and light pollution by a large percentage. Of course, a certain percentage of light is reflected off the ground surface back into the sky even with full cutoff—but that percentage is especially small when the surfaces are dark.

We all want roads safe to drive on and neighborhoods safe from crime. But glaring lights in drivers' faces and all over our properties are detrimental to these aims. When you drive down the street you should see only the illuminated path before you, not the bright bulb itself. Similarly, it hinders the protection of property to light everything, light too brightly, or have light being cast up and to the sides. By the way, studies by the FBI and other organizations have not established that outdoor lighting really does anything to reduce the incidence of crime. Surely the reason is partly that so much outdoor lighting is poorly distributed or overdone (consider that in many communities more burglaries are committed in broad daylight than at night!). If an area is not properly patrolled or otherwise protected, the presence of a lot of light may actually help criminals. To whatever degree

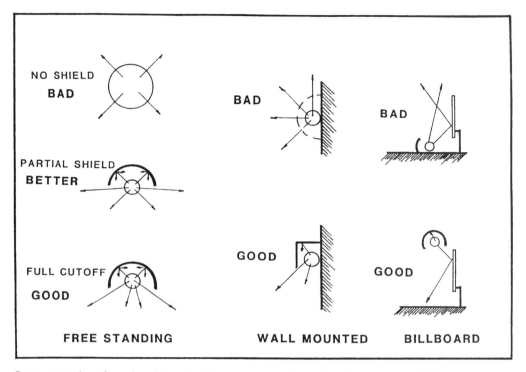

Some examples of good and bad shielding. Courtesy of Dr. David Crawford, KPNO.

proper, well-directed lighting can or cannot discourage crime, the fact remains that we want the light in our streets and key points on our property or home, not in our eyes and the sky or everywhere. And that is what good shielding and placement are for.

Principle #2, using outdoor lighting only *when* it is needed, is exemplified by ad lighting curfews. The light pollution from Tucson, Arizona, decreases by about 35 percent between 8:00 P.M. and 1:00 A.M. because of compliance with such a law. The city of San Diego and San Diego County in California have also adopted ad lighting curfews. The idea is to install timers to turn off the ad lighting (not the security lighting) of all businesses not open to the public after a certain hour. In all but very large cities, advertising your business at 4 A.M. is not a very effective practice. More money is likely to be wasted on the electricity than is gained in sales. But even in cases where this may not be true, and where the cost of the extra hours of electricity is considered affordable by the business, the total demand for electricity drains our resources and generally ends up raising the bills of all of us. In Arizona, important business leaders have stated categorically that outdoor lighting reforms have not had adverse effects on their sales.

Principle #3 says that we should use always the best *kind* of lighting for a particular lighting task, and for many tasks that means low-pressure sodium (LPS), the most cost-effective of all.

Incandescent lights and mercury vapor lights are really quite inefficient compared with LPS and high-pressure sodium (HPS) lights. Between HPS and LPS, professional astronomers and many quality lighting experts favor LPS, and for more reasons than just its lower cost. One reason is that LPS provides the greatest visibility per amount of illumination. The U.S. Navy uses LPS on the flight decks of its aircraft carriers. Many other federal installations, as well as chains of supermarkets and department stores, have opted for LPS for visibility in security lighting. The reason the eye can see with greatest acuity in LPS is that this light is almost purely monochromatic (of one wavelength or color), and the wavelength is near that at which the eye is most sensitive. The monochromatic light of LPS happens to be the

Relative efficiency of different kinds of lighting.

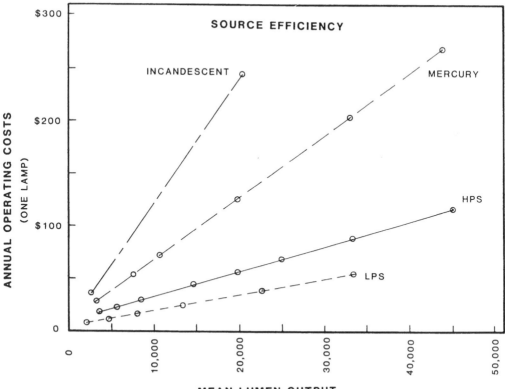

cause of its one great drawback, though: it cannot render colors properly at all. Thus some law-enforcement officials have objected that the color of cars or the clothing of people fleeing from the scene of a crime cannot be determined. In nationwide polling, however, most law-enforcement officials apparently felt that the greater visibility provided by LPS easily made up for the color problem. Certainly for indoor security lighting it is best. Outdoors for security or for near intersections (where the color of traffic lights needs to be recognizable) the answer can often be the use of a small amount of other light in addition to LPS. Obviously, high visibility is what is needed on streets. A number of cities (Houston, for instance) have made LPS the lighting of choice for their highway systems.

Professional astronomers have their own special reason for advocating widespread use of LPS rather than HPS or other kinds of lighting. Their problem with the other kinds, certainly with HPS, is that they radiate at so many different wavelengths. By using a simple filter, the single wavelength of LPS can be blocked out, leaving astronomers with a still-good view of the universe. But HPS radiates at too many wavelengths, including around many at which important celestial sources do. The professional astronomers at the famous Palomar Observatory in California devised the "picket fence" analogy to describe the situation: LPS is like a single board they can see around, but HPS is an almost continuous series of boards making up a fence that blocks almost all of their view. The Palomar astronomers eventually convinced San Diego and San Diego County to use LPS in their streetlighting systems (as well as to enact other anti–light pollution measures). The city did not want the negative publicity of having helped ruin the effectiveness of the world's most powerful and discovery-making observatory. But the city council would never have gone for LPS instead of HPS if there had not been economic and other practical reasons, plus the support of newspapers and citizens for the choice. Another instance of conversion to LPS associated with a professional observatory but finally made for economic more than astronomical reasons is that of San Jose, California. Like San Diego, San Jose is saving *several million dollars a year* by the conversion to LPS. The San Jose report indicates that astronomy at fairly nearby Lick Observatory was only one factor, and not the dominant one, in the decision to make the conversion. The astronomers at Lick are pleased, nonetheless, because the effect they have gotten is decreased sky-glow from the city. The Lick astronomers have been able to make observations on very faint objects that light pollution had prevented for many years. Light pollution *can* be reduced!

Is LPS something for amateur astronomers to push for in their own areas, even if no professional observatories exist there? By all means. The savings will be great because the amount of light needed to produce a given visibility will be less. It is crucial, however, to stress one point: conversion (from incandescent or mercury vapor or HPS) to LPS should not be to needlessly increased illumination and therefore minimal savings but rather to equal (or, if warranted, reduced) illumination and huge savings. In all too many cities, the discovery that HPS or LPS is so much cheaper than mercury vapor has led to a needless increase in light due to the thinking that more is always better. Bear in mind that numerous high-wattage, high-illumination LPS lights—especially if poorly shielded—will still waste money and still fill the heavens with their glare so that the beauties of the starry sky are lost. Some people have dreamed of a day when cities will be mostly lit by LPS and special glasses that filter out this wavelength will be cheaply available. You would put on the glasses, and even in the midst of a large city a starry sky would spring into view. Although this would be a good idea, of course, it is no full solution in itself. We should all work for other lighting reforms too so that even without glasses more stars will add their touch of splendor and inspiration to city skies.

And not so many miles away from urban centers no special glasses will be needed to see the starry sky in its full glory as it should be.

SHOULDN'T GLAMOROUS CITIES BE BRIGHT?

There are people who argue that cities should be ablaze with lights, that this is part of their glamour. Well, there is indeed a genuine beauty to the sight of thousands of city lights twinkling ahead of you, or below you as you approach by plane—providing, of course, that you do not think about the dangerous waste that is what you are really seeing. When you consider what these lights are like close up, too, there is certainly not much beauty left. That great scholar of language and writer of fantasy J. R. R. Tolkien picked the streetlight as a typical "product of the Robot Age, that combines elaboration and ingenuity of means with ugliness, and (often) with inferiority of result." He was certainly right about the inferior results from most of our poorly designed (or poorly used and placed) streetlights right up until today. But of course even if we take the care we should (and must) in better design and placement of street lights, they remain mass-produced objects that are essentially functional, destined to be replaced and updated, and—most damaging of all to any pretense of beauty or elemental interest— entirely understood in this, their limited role, by us, their makers. This

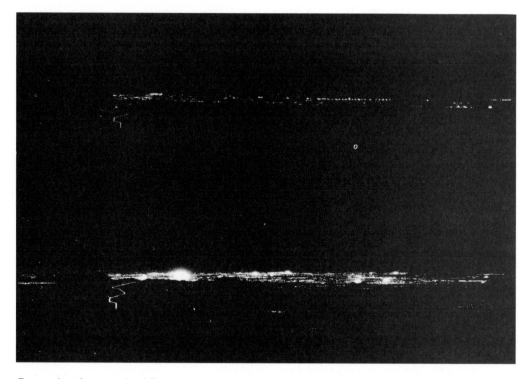

Composite photograph of Tucson, Arizona from Kitt Peak National Observatory. Upper portion photographed in 1959, lower in 1980. Courtesy of Dr. David Crawford, KPNO.

point is the answer to anyone who would argue that, close up, stars are merely giant hulks of hydrogen and helium (in many cases, and for the most part). Stars are far more than that! They are among the most important units in the structure of the universe of matter and energy, time and space, but it might be more accurate to say that these complex, independently existing, more-than-functional objects are characters in the story of organisms (nonbiological though they be) in the life of the universe—and not just the universe of matter and energy, but also the universe of interest and beauty. They are both revealing and mysterious in greater amount and in many more modes and levels than merely functional products of mankind could ever be.

As Tolkien says, there are more permanent and fundamental things in the universe than streetlights—like lightning, to follow the example he gives. Tolkien scoffs at those who feel that things like streetlights are more a part of "real life" than stars or birds or even, in one sense, than the strongest inventions (findings?) of man's fantasy:

> The notion that motor-cars are more "alive" than, say, centaurs or dragons is curious; that they are more "real" than, say, horses is pathetically absurd. How real, how startlingly alive is a factory chimney compared with an elm-tree: poor obsolete thing, insubstantial dream of an escapist!

And as for the sky and those clouds, so substantial in their insubstantiality:

> For my part, I cannot convince myself that the roof of Bletchley station is more "real" than the clouds. And as an artefact I find it less inspiring than the legendary dome of heaven. The bridge to platform 4 is to me less interesting than Bïfrost guarded by Heimdall with the Gjallarhorn. From the wildness of my heart I cannot exclude the question whether railway-engineers, if they had been brought up on more fantasy, might not have done better with all their abundant means than they commonly do.

To Tolkien, fantasy involved strange, mythic forms clothing (to make more vivid and wondrous) the essence of fundamental things we all need to keep seeing afresh: like love (for many things), and evil, and lightning, and the sky, and the stars. But a science like astronomy keeps revealing to us such fresh and remarkably unusual visions that it fulfills much of the function of fantasy for its enjoyers. And, of course, it does this with the involvement of technological instruments and verifiable experiments that can still the unease of those of us who are not willing to accept even for artistic purposes the invention of a secondary, fantasy world. In light of this aspect of astronomy, we could say that perhaps railroad-engineers—or anyone engaged in any "practical" business—might do better if they had been brought up with more exposure to the beauty and strangeness and vast scope and room for bold surmise that is the heart of astronomy.

Speaking of bold: I dare to make such a flight into the rarefied atmosphere of philosophy (about stars and streetlights) because I write here for an audience already favoring the stars. (You are unlikely to be poking into this part of this book unless you have already read about and in some measure been sympathetic to the astronomical wonders which are its topics.) But let us return to that question of the beauty of city lights, and their glamour, with a more practical point: cities with fabulous monuments or world-famous entertainment districts are surely not going to dim the dazzle of their casinos and the like, which bring them tourist dollars. Yet even here an excellent compromise can be made between stars and modern city life: if cities want to highlight these areas, what better way than to limit extravagant, bright, showy lighting to them alone? Most of a big city can be lit safely and efficiently with modest, shielded (often LPS) lights. Save whatever glitter and glamour lights can have (and it does not always have to be blindingly bright or pointed skyward!) for your showpieces. Let New York's Great White Way be all the more a proclamation of the good things it holds by its standing out from the quiet, efficient, and money-saving lights used for most of the city.

DARK SKIES FOR COMET HALLEY

How does one publicize the three principles of good outdoor lighting, and such things as full-cutoff shielding, ad lighting curfews, and LPS lights, which embody them? Talk with friends, with people who know politicians, with fellow astronomy club members certainly. The letter to the editor of your local newspaper, and to certain politicians, remains something that can work. But I found myself involved in a very special campaign to publicize light pollution and its cures a few years ago. The results were amazing.

I founded and directed the project for five years. With the direct effort of only myself and a very few friends and fellow amateur astronomers, I managed to bring the message about light pollution to perhaps many times more people than had ever heard it before, all over the world, in some of the world's most prestigious newspapers, in several languages, to people of ages from six to sixty and much older. But I cannot take the primary credit. That, I think, belongs to the entity which got the world to listen in the first place: Halley's Comet.

It was in the spring of 1981 that I started thinking more seriously about the fact that light pollution would deny millions of people their chance to see Halley's Comet in 1985–86. And so it was that I founded Dark Skies for Comet Halley (DSCH). The ostensible goal of the project was to get many cities to turn off their nonessential outdoor lighting for a few hours on the best nights of the comet. The second, even more important, goal was to publicize the need for permanent reforms in the use of outdoor lighting. This second goal seemed more possible than the first, but many people were skeptical that the media and general public would listen. Some of those in positions of power in the astronomical world said they would be willing to help—but only if and when I had proven I could get wide publicity! In contrast to such fair-weather "friends" were those like the astronomy author Guy Ottewell and the leaders of the large confederation of American amateur astronomy clubs called the Astronomical League. Not only Guy but also League representatives like Don Archer, Jerry Sherlin, and Carol Beamon (later Tom Martinez and George Ellis) were supportive from my very first contacts with them about DSCH and in some cases committed much time and energy to helping make it work. Professional astronomers like Peter Boyce and Ray Newburn Jr. were also encouraging to me even in the early stages. Later, individual amateur astronomers like Rick Kurczewski (especially Rick) heard about DSCH and helped. Indeed, one of the major accomplishments of the project was identifying people all over the country (and world) who had been trying in their own way to fight light pollution and now could know that their efforts might eventually be unified.

I could not have gotten very far with DSCH, though, if it were not for the help of the person who has probably done more than any other to study the light pollution problem and help spread the gospel of dark skies and quality lighting. That person, instrumental in many of Arizona's victories against light pollution, is Kitt Peak National Observatory astronomer David Crawford.

As it turned out, my success on the first DSCH goal—actual light dimming for Halley—was minor. We came very, very close to some stunning successes, though. In answer to my request, New Jersey State Senator James R. Hurley helped sponsor a DSCH bill which passed the State Senate unanimously but did not come up for consideration in the Assembly in time. The bill required the state government to request all New Jersey municipalities to try light dimming for Halley and for those municipalities to receive an information package on Halley and light pollution. As the climax of Halley drew near, Senator Hurley and I changed tactics and sent out (at his expense) joint letters to the more than 500 municipalities in New Jersey. Only a few percent wrote back requesting information, but these were generally quite positive and interested, and I believe we opened some important doors for permanent reforms in the future.

But the most spectacular case of DSCH nearly succeeding was in what a space shuttle astronaut affirmed as the brightest city on Earth as seen from orbit: New York! Mayor Ed Koch was apparently influenced by DSCH (though perhaps indirectly through *The New York Times*?) and on Thanksgiving Day 1985 called for city-wide dimming of nonessential lights. It was a front-page article in *The New York Times* the next day, and both my name and comments were featured prominently! A single amateur astronomer with hardly any money and no particular civic connections had gotten himself mentioned in major national news and in the counsels of some of the nation's most well-known politicians—with help from a handful of other devoted amateurs and the mighty ally called Halley's Comet. I would have

traded that considerable personal thrill in a second, though, to have gotten Koch's initial wishes realized. Unfortunately, although his science advisors apparently felt that a city-wide dimming of nonessential lights was quite feasible (as it surely was), Koch retreated and settled for the hopelessly ineffective alternative of turning off virtually all lights in a few well-policed parks. That did not significantly reduce the sky brightness with miles of city all around at full glare. But at least it did gather thousands of people who might otherwise have missed Halley to see it through the expertly manned telescopes set up in the parks that evening in January 1986 (before dawn in March would almost certainly have been better, though). And why did Koch decide against the effective, city-wide plan? His chief science advisor told me it was probably out of "political considerations."

The tremendous success of DSCH came in its second and more important task: publicizing light pollution and its curability. My ordinary life was almost overturned by thousands of letters from interested people and by national newspaper and radio-show interviews that were live on the phone. (One of these, I remember, woke my wife and me in the early morning with a one-minute preliminary warning that I was going on live via the phone onto a major city radio program—if that was all right with me, of course!) The Astronomical League published my quarterly newsletter, *DSCH Journal,* for more than 9,000 readers, including ones from every continent in the world except Africa and Antarctica. There was even a Spanish translation of it distributed by an organization in Buenos Aires. DSCH narrowly missed out making it on the best Halley TV special and into *Newsweek*'s cover article on the comet. But I was interviewed for ABC national news radio, and dozens of major newspapers and magazines mentioned DSCH. (I got more than 400 letters—mostly from different schools—in little more than 4 weeks just from kids who read about DSCH in *Highlights for Children* magazine!)

As Halley's Comet recedes, even the message that most people missed it because of light pollution is becoming less effective. A new possible source of publicity around which to rally efforts is the European Space Agency's awful plan to launch in 1989 a reflective space ring that would be so large that it would have a wider apparent diameter than the moon. Its purpose would be to commemorate the centennial of the Eiffel Tower; its result would be to wreak havoc with the work of professional astronomers and set the precedent for propaganda and advertising in space—how long before political ideologies and Coke and Pepsi signs are crowding the heavens? The ESA "space ring" should stir a controversy that involves many important issues, but one of them is light pollution here on Earth.

If the "space ring" gets into orbit, it will be about as bright as Jupiter and so would be a more visible topic for discussion than Halley's Comet. But it would be an ally only in the sense that it may raise a righteous uproar to protest it.

HOW YOU CAN HELP DEFEAT LIGHT POLLUTION

All of this talk of the principles of good outdoor lighting and publicity boils down at last to the crucial question: how can the average person get the lighting reforms considered and passed in his or her area and thereby reduce light pollution?

Contacting a local politician is likely to be effective only if you know how to go about it and have impressive, documented data to give him or her. The place to gain this know-how and data should be one place for all opponents of light pollution, and it looks as though we are going to have it: a central bureau for light pollution information.

As I first wrote these words in early 1987, David Crawford and his associate William Robinson had several small computers for data, a light-pollution slide show, a planned videotape, and lots of written information. Some of the written information was collected in a 1985 booklet by Crawford that should still be available, like these other items, at cost. Now at my last edit in 1988, Crawford welcomes donations (and members) to the new nonprofit International Dark-Sky Association (IDA). Before, efforts were handicapped by dearth of funds. By the time you read these words of mine I am hopeful that the IDA will have received enough publicity to have upgraded its services with incoming funds from supporters who can spare the money for this excellent cause. In my opinion, opponents of light pollution must unify, and the IDA is the best place to do it. There are amateur astronomers who spend many thousands of dollars on telescopic equipment, and they had better start realizing that sending some of their money to help battle light pollution is not just a rightful payback for the enjoyment they have gotten out of astronomy—it is a wise, indeed a necessary, investment if they are to go on observing. No amount of money is going to buy a large enough telescope or effective enough filters to preserve their view of the sky indefinitely unless light pollution can be minimized.

Before, individual amateur astronomers and individual astronomy clubs were fighting alone. Now, with the institution of IDA, we will have access to written, audio, and visual information which demonstrates beyond any doubt why we need outdoor lighting reforms, how they are intended to work, and how they *have* worked—in cities ranging from the smallest all the way up to just about the largest (containing populations of more than a million). The IDA should also provide specific advice to each individual who wants to proceed, or wants to tell activist friends how to proceed, in his or her own area.

I cannot predict exactly to what state the organization will have progressed by whatever time you read these words. But you can already receive possibly all the information and advice you will need by writing to:

Dr. David Crawford
International Dark-Sky Association*
3225 N. First Avenue
Tucson, AZ 85719
www.darksky.org

In my own home region, I have recently persuaded State Senator James R. Hurley to propose a bill that would form a light pollution study commission in the state of New Jersey.

If we can get such a bill passed in New Jersey, it will be the most powerful precedent ever in the battle against light pollution. I hope you will be

*2002 update: IDA now has roughly 10,000 members in about 70 countries!

learning that progress has gone well on it when you write to Dr. Crawford and the IDA.

CONCLUSION

In 1986's last issue of *Time* magazine, the cover article was in the form of a letter to the people of one hundred years in the future. It is a little surprising, and certainly a little sad, that a person writing in 1986 should choose one century ahead rather than seventy-five years—to the next return of Halley's Comet. But the article, written by Roger Rosenblatt, is a brilliant and beautiful piece, with many piercing and often hopeful observations. *Time* magazine should be congratulated for choosing to feature, in large print on its cover, the final words of the essay. Here, addressed to our children's children's children are the words: "Do you see starlight? So do we. Smell the fire? We do too. Draw close. Let us tell each other a story."

Again the appeal to starlight! But *Time* and Rosenblatt, in the midst of our commendations, must also be criticized—with so many of the rest of us. They, and we, have not been looking at that starlight recently. If they had tried they would have found it already missing from some of America, struggling in death-throes over most of America, still shining in its full heart-filling and heart-breaking beauty over precious little of America. The people of the year 2086 will most certainly *not* see the starlight—unless we do something about light pollution.

The reasons we have not been looking at starlight ourselves lately are in the self-devouring circle of ourselves and the environment we have made around us. There is a callousness, wastefulness, and foolishness that let the stars fade in our costly glare, but then those weakened and uglied skies in turn form an environment which makes possible—almost inevitable—worse callousness, wastefulness, and foolishness. George Orwell wrote that the English language becomes ugly and inaccurate when our thoughts are foolish, but the sloppiness of the language makes it easier for us to think foolish thoughts. He also said that using empty catch-phrases and excessive words and letting well-worn expressions write—and pervert—our language for us was like anesthetizing whole areas of our mind. Perhaps he should have been more forceful! We should, at least, be more forceful about light pollution. Losing the stars does not just anesthetize parts of our mind; it kills them, as does cheapening our language. To dilute or cheapen our words is to destroy parts of our mind and our morals, and also to make the world we see turn gray and barren and trite. The turning of our world into a waste may be partly a consequence of poor speech, but it is not a mere figure of speech. You do not have to look far to see that the unlivable deserts of many minds have swept across parts of the outer world we share, destroying their life and beauty for all of us. And the circle goes around: drain the variety and beauty out of a person's living environment, poison it or pollute it, and you eventually do the same to his or her mind.

Do you require further incentive than the desperate worldwide need to conserve energy, the billions of dollars we can save ourselves, and the majesty of the stars restored to the rest of your life? Then think about the stars fading entirely out of the eyes—and hearts—of our children.

The stars have given their constellation patterns to some of our most poignant and true tales of life. Their beauty has inspired and helped lift us out of what might otherwise be despair to a higher, grander, but also more piercingly clear view. If the stars are driven from our skies, what precious parts of our minds and spirits that they reflect and encourage will be lost from humanity as well? A world without stars—to Dante that meant Hell. If we let our wastes hide the stars completely, we are committing another important act in the building of such a hell around ourselves and our children. To alter slightly what Guy Ottewell has said, light pollution is not so dangerous an evil as are nuclear buildup, ecological destruction, and a few of the other great problems of our age—"yet the problems of an age are often solved in clusters, since they arise from related causes." If we let the stars be lost, we are cutting from humanity's mind one of the very sights that has most inspired poet and scientist, sage and lover, man and woman, young and old since humanity became human.

We must not let it happen.

13

The Best and Worst
of Returns

IT WAS THE BEST of returns; it was the worst of returns.

The 1985–86 apparition of Halley's Comet—*our* return—was both best and worst in more ways than any before. Not just increased technology like Halley-visiting spacecraft and Halley-obscuring light pollution caused superlatives. So did increased physical knowledge and observational knowhow. Many more bests and worsts were due to the most distinctive choreography of Earth, Sun, and Halley in 3,538 years of 48 calculable returns from 1404 B.C. to A.D. 2134.

Try working out the equation of bests and worsts to determine how good or bad this return was overall. The return with first spacecraft visits to Halley, with an unusual *two* moderately close approaches of the comet, with the longest Halley naked-eye visibility ever, with the largest Halley true head-size and Halley pre-perihelion brightening ever properly recorded—all this occurred for the most well-informed, optically well-equipped, and simply largest potential audience the comet has ever had. But then there is the negative side of the equation. The return with Halley's poorest maximum brightness since maybe 986 B.C. (when King David probably missed it), with the comet close when less active and far when more active, with the dust tail angled to look long only when far enough and still spread enough to appear extremely dim—all this occurred for the most distracted, by far most city-light-polluted audience ever.

Which side won? I think there are two different answers for two different groups of people. But before I give them, let us review—with impossible brevity—the nine-month main show itself.

• • •

Most serious sky watchers with large backyard telescopes first glimpsed Halley late in the summer of 1985.

For me it came in the cold country quiet of 4 A.M. on September 14 near where a quarter-century earlier I first stood in wonder beneath the stars at age six—and now stood seeing through a thirteen-inch telescope a 13th-magnitude patch of light in Orion's club and under the main asteroid belt (farther out than anyone first spotted it in 1835) just a few degrees from Comet Giacobini-Zinner. It was Halley's Comet. I had lived to see it, the generation-bridger, the once-in-a-life light. No one could ever take that away from me.

And so came Halley Autumn. Halley exceeded all forecasts of brightness and size, both, in some weeks, doubling! Authorities said Thanksgiving's magnitude-5½ Halley might reach a prominent 3rd magnitude by Christmas and peak at spectacular zero magnitude a few months later. Alas, that it was not to be! According to Charles Morris of the Jet Propulsion Laboratory, there was no anomalous autumn outburst, just Halley's typical behavior, which the narrow-field telescope users at past returns had not recognized fully or recorded properly for us. The outer coma of this particular comet becomes bright enough to see with impressive rapidity, and the increased size leads in turn to still more increased brightness—for a while. In October 1985, fading Comet G-Z was replaced by Hartley-Good and Thiele, two more herald or vanguard-of-Halley comets bright enough for binoculars and still brighter than Halley but lacking that chillingly significant jewel in its head—the star-like "false nucleus" that may have been the most intense of our lives for a comet this far out. November 1985 was all-night, first close-approach, Mars-orbit-crossing, Pleiades-passing, head-on, and thus widely diverging branched tail Halley—first seen with naked eye by Steve Edberg and Charles Morris on November 8.

In 1½ months after early December 1985, Halley brightened only from magnitude 4.8 to 4.3 as both its apparent and true size shrank because of increasing distance from Earth and decreasing distance from the sun (now close enough to have coma-compressingly strong solar wind). In early December the visible coma (head cloud) reached largest apparent and true diameter for the naked eye—more than ½° and a million miles for many observers. A maximum of 1½° and roughly 3 million miles (the latter by far an all-time record for any comet) was reported! Ultraviolet imagery later revealed the invisible hydrogen coma of Halley to be much bigger than even Comet Bennett's in 1970, in fact 21 million by 15 million miles (large enough to almost fill the gap between the orbits of Earth and Venus).

Quite visible in December were exquisite ion rays in the several degrees of gas tail which itself in early January kinked, disconnected, and distorted (mostly on photographs). And on January evenings (also March mornings) Halley's Comet became the quest-object for crowds of thousands everywhere, as many as 40,000 at Jones Beach, New York, on January 11, 1986. Few DSCH (Dark Skies for Comet Halley) light-dimming programs were tried, but the problem of light pollution was tremendously publicized, and many doors toward permanent reforms opened thanks to the ally Halley.

After about two weeks' unviewability around February 9's perihelion (mere days after the comet was rounding the far side of the sun from us), Halley's Comet reemerged before dawn. Truth now accepted: the comet would not be a spectacle. In the Southern Hemisphere, though, the gas tail was traced to a maximum of about 18° and 0.4 A.U. about 37 million miles long (for most northerners, only up to about 6°). The now magnitude 2½ comet showed at first some gold in head and startling aqua-blue in gas tail, and a very few people saw more than the stub of the huge, too widely spread, bent back dust tail. Countless children will recall all their lives the delicious strangeness of those 3 A.M. wake-up calls for a drive to family adventure. And good telescopes showed a blue head with planet-sharp inner coma and intricate structure as splendid as the Great Orion Nebula's. At high magnifications some observers were able to see and sketch still more intricate jets from the comet's heart.

In March 1986 two spacecraft passed near Halley's head and three right through it, the trio suffering extensive damage from dust bombardment but radioing back most of their data. The two *Vegas* and one *Giotto* spacecraft were from the U.S.S.R. and Europe, respectively, but really from a whole world never before joined so well in cooperation on a space mission. They sent back treasures. These included photographs of a somewhat hilly, ellipsoidal nucleus about 10 miles by 5 miles (much larger than most scientists had thought) and composed of ice (vindicating Fred Whipple's 36-year-old "dirty snowball" theory) but black as coal (from what is probably organic—but almost surely not living—material on the surface) and far hotter-surfaced than imagined (135°F. in one measurement). The big, irregular shaped, hot, black, lovely heart of Halley's Comet was spinning around once every 52½ hours and spouting like a whale but up to 7 immense bright jets at a time, gushing typically 3 tons per second of dust and variously 21 to more than 60 tons per second of gas that was 80 percent water vapor.

But further study of the spacecraft and Earth-based professional data revealed still greater surprises and more moving beauties. The nucleus of Halley's Comet is so porous it must be somewhere between 33 percent and 90 percent empty space—similar in this respect to freshly fallen snow of

The active nucleus of Halley's Comet

particularly wet, "heavy" variety. But while the nucleus is as porous as a snowdrift, it is still certainly as brittle as ice—the strange combination a result of the cold and vacuum of its birthing place far from the sun. The nature of Halley's nucleus throws new light on where and how the comets have formed and also on the raging controversy about whether collisions with comets or meteoroids have caused the great (possibly periodic) mass extinctions like that of the dinosaurs (even just the larger size and larger mass, despite far lower density, of Halley's nucleus may have a great effect on changing some of these theories). Other unexpected discoveries include the fact that Halley's dust comes in so many sizes (and in especially great quantities of very small particles) and the composition of the ice: not surprisingly, 80 percent frozen water but rather surprisingly only 3 percent or less carbon dioxide and about 17 percent frozen carbon monoxide (the last of these substances is responsible for certain especially powerful bright jets and hence is presumably in some places concentrated in rich clumps in the nucleus).

Numerous other findings will keep scientists busy for years to come (especially as they plan for the exciting proposed mission that could send a spacecraft, more slowly than the Halley ones, past three comets and two asteroids between 1994 and 2005). But one final discovery about Halley's nucleus merits special mention: the way it spins and tumbles. The true rotation period is 7.4 days around the long axis of the nucleus—the same way a bullet or a well-thrown "spiral" pass in football spins. But the 52½-hour (2.2-day) period is a "precession" about an axis inclined almost at a

"3-D" diagram of Halley nucleus.

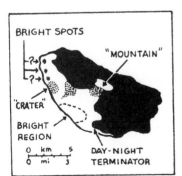

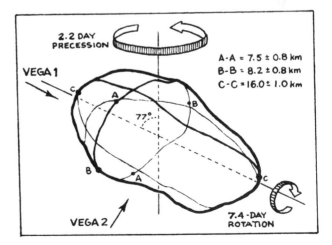

right angle (77°) from the rotation axis. The nucleus is thus moving through space like a supremely ill-thrown pass in football. Could anyone throw a pass so beautifully bad? Any fairly small object (certainly a bullet or football) would move this way in space without the help of aerodynamic forces to keep it stabilized. And that includes the Manhattan-size bullet or football that is the nucleus of Halley's Comet.

After the spacecraft pilgrims visited the comet in March, thousands of Earth-bound human Halley pilgrims in early April saw the gas tail dwindle just as they reached the Southern Hemisphere. The all-night and overhead comet looked like an eerie triangle with short gas and short broad stump of dust tail. Besides exotic lands and still more exotic skies, some of the journeyers got to see one of Halley's largest (but not very large) flares— perhaps more than half a magnitude up to the return's maximum brightness of about 1.9 on April 6.

Two full weeks after April 10's very southerly closest approach came perhaps the return's biggest surprise for observers. While veiled by moon-light, the dust tail narrowed threefold because of our changing perspective as the comet fled out and down. With the tail's increase in apparent surface brightness (we saw through a greater thickness or cross-section), *presto!* it appeared, though still very dimly, in all of its immensity. During the total lunar eclipse of April 24–25, the surprise was first fully unsprung, with Terry Lovejoy in Australia tracing about 43° of tail! In the next week or so of moon-free evenings, other observers (including myself in New Jersey) may have glimpsed even more. I saw a length of 20° to 30° on several nights (the tail positions and configurations agreeing well with those sketched by Charles Morris), but whether I really saw 50° in the best sky of the year (was my naked eye glimpsing chains of very faint stars?) remains uncertain. I have even heard thirdhand rumor that someone in Australia saw more than 60° of tail with its end lost in the Milky Way! Such claims are difficult to prove. But whatever the maximum length seen, the eye may have done better in this instance than any form of photography—the handful of de-voted and persevering observers who glimpsed, alone of all people on Earth, the long tail got what they deserved (though they had to have the luck of superb weather that week to get it).

Morris's models and calculations suggest that the viewing angle on the curved dust tail was so good in late April that even a 60° tail would be "only" about 0.8 A.U. or about 75 million miles long (Morris does not support a 60°-long tail, however, merely recognizing it as a possibility). Superb viewing angles (often with the comet closer to Earth and sun) are a key to Halley's great angular lengths at many returns.

Those late-April and early-May evenings there was the feeling of a giant's form or shadow passing just by and glimpsed by only a few alert and shaken watchers while the rest of the world was sleeping or looking the other way!

After the long tail, the rest of the return was anticlimactic save for how long Halley would remain visible to the naked eye. Reports of naked-eye sightings extend through May 30, 1986, but Halley was near a 6th-magnitude star the last few nights, so positive identification was difficult. In early June the comet may have flared a bit, and a surprising anti-tail first seen in May was still visible. The comet was a huge diffuse object, finally lost in evening twilight after Richard Fleet's final observation with a 14-inch reflector on August 9 in Zimbabwe.

So Halley held surprises to the very end of this main apparition—and remained interesting when reemerging as a difficult, diffuse object of from 11th to about 14th magnitude in the largest amateur telescopes from about November 1986 to June 1987. During this period it passed on out beyond the orbit of Jupiter. Maybe the last direct sight, marvelous as amateur Steve O'Meara's 19½-magnitude Halley Hawaiian mountaintop gaze in January 1985, will come in 1988 or 1989. The Hubble Space Telescope and its successors may keep Halley recorded all the way into the far dark until its next approach in 2061—when today's babes will be silver-haired.

Now to the two different answers about how good this return was for two different groups of people.

Did the public think this return of Halley's Comet good? The public wants spectacle and did not get it, but they had been prepared for the worst, so they thought this *return* a flop but not necessarily the comet. *The New York Times* reported that January observers in the city were very disappointed by the comet's dimness (no wonder in such a light-polluted sky) but often seemed even more thrilled just to have glimpsed such a legend. There was surprisingly little rancor or contemptuous laughter anything like there had been for 1973–74's much-ballyhooed and then supremely disappointing Comet Kohoutek. Wonderfully, there was also almost no superstition or injury from mayhem. The comet seems to have escaped (even after having been T-shirted, Frisbeed, and written about in close to one hundred books) with most of its legend still looming large in the past and future. A Gallup poll said that about 6 percent of adult Americans claimed to have seen it even by January 1986—but with so long a time visible and so much information on how to find it (a whopping 92 percent knew that Halley was back), maybe ten times as many could have seen it but for light pollution.

And what about the assessment of the true comet and Halley devotees? The wonders of this return, the good side, often won out. When they did, however, it was usually the hard way—as with most worthy things. *Many of us bled our blood for this comet.* We were drawn to it exhausted every night (161 successful nights and 10,000 auto miles during the main apparition for Morris!), we labored to explain and locate it for the throngs, we begged our families' forgiveness for the ringing phones and far-off looks and ceaseless comet-talk and odd-hour absences. More privately and crucially, we bled our blood in hopes and heartaches. And all of this most intensely for more than nine months—to what birth? Through it all, through three full seasons and every single hour of the night, through half a hundred Halley guises, there was a steadily accumulating beauty and reward. Those who bleed blood do wish to believe their ordeal justified, but I can honestly say that, wishes aside, it really was for me—and for many of us.

It was the best of returns; it was the worst of returns. And it was *our* return—especially those of us who sought Halley with devotion. For us—disappointing and dim as this return often was—it was even more than a passage of the Comet of All Lifetimes. Whether dim or bright, short or long, this comet is something that penetrates all areas of human life, something far more and deeper than the purely astronomical. It is a beauty unbidden because far more difficultly lovely and disturbing than what we would dare to ask for. It is a fruitful frustration from which we cannot be freed, a riddle (not just of science but of ourselves) of which we cannot be rid. Those of us who look back know that Halley's Comet, like life itself, was a trouble and a treasure.

2134 A.D.: A manned spaceship approaches Halley's Comet

14

The Fate and Meaning of Halley's Comet

AT ALL RETURNS of Halley's Comet before 1985–86, people thought the comet's appearance a fateful event. Multitudes believed that a ruler or a nation would be influenced (usually afflicted) by the comet. In 1910 Halley's tail came close, and many people decided that that most beautiful and eerie part of the comet would brush us all with silent, invisible death: humanity, born of Earth, would perish by a comet.

The worries which many of us now have at the end of the twentieth century are different. The possibility exists that mankind will destroy itself, and thus the 1985–86 visit of mankind's comet be its last ever observed— by us, at least. The comet would go on gliding around its stable orbit, sweeping its tail through Earth-orbit-sized arc of space and burning brightly across the inner solar system once about every seventy-five years—what had once been called a human lifetime. But the Comet of All Lifetimes would have no more lifetimes to measure or to kindle with wonder.

Yet another prospect exists, necessarily less disastrous than the previous one from our standpoint, but perhaps as much a signal of our inner destruction, if not outer. And that is the prospect of humanity in some manner purposely destroying or depleting or degrading Halley's Comet. A human race that committed such an act might have lost all vision of what we have tried to mean by the word *human*.

In recent years comet scientists have come to the conclusion that Halley's Comet will probably not depart markedly from its present orbit, or exhaust its ices and thus its brilliance, or break to pieces from any kind of natural cause, for what is very likely to be thousands of years at least. Yet surely we are few Halley-periods away, probably less than one, from the time when mankind could have ample opportunity to spoil or even utterly destroy the comet in a number of ways. Why anyone would want to do this can be learned from studying the motives and character of people who have already performed or permitted similar acts of many of Earth's most beautiful and meaningful places and things and creatures—not to mention upon other human beings.

But if Halley's Comet really could be ruined by mankind at its next visit, then now, before the 1985–86 visit has faded too far into memory, is the time to think about that possibility: and by thinking, speaking, and acting about it, to help prevent its ever occurring. To say that the fate of Halley's Comet is now forseeably within our hands (if we do not destroy ourselves) must absolutely not be a boast of power, for that kind of power is only a physical ability to exercise empty destruction. Halley's Comet can truly be ours in only one sense, be ours only to whatever extent we can understand and appreciate it as both reality and symbol in our vision of the universe. There is possibly no end to how much more Halley's Comet can become ours in this sense. Our achieving of the dangerous ability to wreck the comet, to very literally control its fate, may be an ill occasion, but occasion it is to ponder more seriously what is the meaning of this life that is in our hands. The life referred to is that of the comet, but because Halley's is distinctively and pre-eminently mankind's comet, that inquiry is into the meaning of our own life as well. Can we find a ray of hope, perhaps like that of a comet's tail sticking up lonely from the horizon of the future, in this search and in what we can do?

First, let us consider more precisely what constructive or destructive things we can do to Halley's Comet in the next few Halley periods.

As Loren Eiseley was growing old, he asked an astronomer friend, hopefully, how close "the great wild satellite" Halley's Comet was to re-turning. Like so many other people of his age, Eiseley had been held up to see the comet as a child in 1910 by his father and hoped he could live to see it again, in large part as an act of love and commemoration for his father. Eiseley was dismayed to learn how many years there were until the next return in 1985–86 and feared—correctly—that he would not make it. But in all the years between 1911 and 1982 (when the comet was first glimpsed on its way back, with electronic imaging), the belief in the comet's return and even in the continuance of its existence was, however probable the validity, one of blind faith. In contrast, when the Edwin Hubble Space Telescope is at last launched (maybe in 1989) we will probably ensure that Halley's Comet will never again be lost from view. Even if the limit of this instrument were 27th magnitude (perhaps better can be expected), that light-gathering power ought to be sufficient to permit a view of Halley until some still more powerful telescopes take over the task and are able to identify the comet's nucleus even when it is more faint, even at aphelion beyond Neptune. So Halley's Comet will never again pass beyond our technically augmented perception, even though the evidence of our senses for its existence and position will rest no doubt at several removes from the live sight through a telescopic eyepiece.

The next step in bringing the comet into our ken beyond seeing it ever better is visiting it with unmanned spacecraft ever better. The first flock of such spacecraft already have paid a short call. Since even a collision of the *Giotto* spacecraft with the nucleus would have had negligible effect, there was no reason to fear any disruption of Halley by humans at the 1985–86 visit. But what will be the advances in unmanned space probes in the next seventy-five years? Surely every planet except possibly Pluto will have been visited and orbited by probes—assuming again, of course, that civilization on Earth does not suffer a calamitous setback or destruction. Although a

Three-dimensional orbital diagram of 2061 return. Derived from *Mankind's Comet*, by Guy Ottewell and Fred Schaaf.

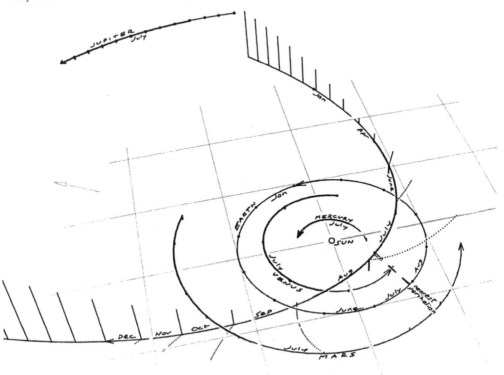

probe sent to catch up to the comet and observe it at its next aphelion is probably quite unlikely, there would be a strong incentive to get one to the nucleus five to ten years before the 2061 perihelion passage—to guarantee our observing early important changes in the nucleus as it approached the distances of Saturn and Jupiter from the sun. Even with instruments like the Hubble Space Telescope and its successors spotting long-period comets early and far out, there might not be sufficient time and fame to get the budget and plans for a spacecraft visit to any comet beyond Saturn except Halley by the middle of the twenty-first century.

Men and women may walk on Mars quite early in the next century. A few greatly longer manned (or "personned") missions—to the Jupiter system, for instance—could conceivably be tried before the 2061 return of Halley. In all such prognostications the political climate on Earth is the toughest yet possibly the most important to determine. Despite the uncertainties, we can venture to guess that no spacecraft occupied by people will join Halley's Comet at its next return when it is still beyond the orbit of Jupiter. On the other hand, can there be much doubt that at least some close passages by manned spacecraft will occur at the next return? The technology will exist, the interest also. The only thing that might prevent sophisticated robot probes from being landed on the nucleus surface would be if astronauts landed there—in which case, of course, instruments would be left by them to go on measuring and recording and radioing back to us information about the environment for a long time. Now that the 1986 spacecraft have sent back their data about the comet and nucleus, it is possible to figure much better the hazards of approaching and landing on the nucleus. My personal guess is that human beings will walk on the active nucleus of Halley's Comet at the next return. If not, then surely the 2134 return should see this long-destined meeting. The extreme closeness of the comet's approach to Earth in 2134 will in any case be appropriate to that or some other *decisive* encounter of Halley's Comet and humanity.

If human beings walk on Halley at the next visit, how odd and remarkable it will seem to observers of the comet. To look at our familiar moon in the sky while *Apollo* astronauts bounced or dune-buggied around on its surface was astonishing. Although no more thought-provoking, even stranger perhaps is it going to seem to stare at the fuzzy glowing ball of the comet's head and try to understand that deep in the midst of that luminous fog large enough to lose the Earth, at the very core within the brightest seed of light, a group of men and women are standing, talking, walking, looking, dreaming. What kind of dream would a person have when wrapped so deeply in the dreamlike form of the comet, asleep within the body of Halley's Comet upon its veritable heart? This would be mortal men and

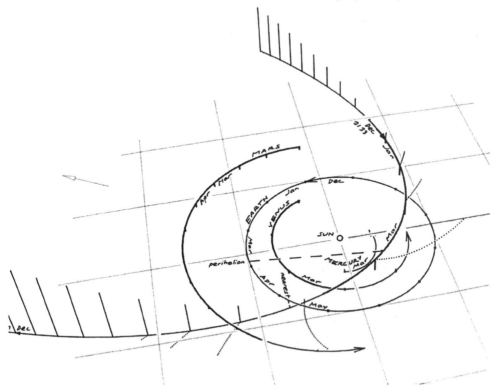

Three-dimensional orbital diagram of 2134 return. Derived from *Mankind's Comet*, by Guy Ottewell and Fred Schaaf.

women making history by being inside that celestial chronicler of history, standing at the source of the beacon of generations, alive in the midst of the latest of the Halley fires to ignite and unite humanity (in awe and inspiration and thoughtfulness, not fear). In one way of thinking, of course, the presence of mundane humans on this most exotic of bodies would end a mystery and dispel part of an enchantment. Yet a first in-person visit to Halley's heart would enter into legend and reveal new mystery. A voyage of exploration, filled with respect to match even burning curiosity, would increase our appreciation of the comet and deepen our conviction that its integrity must be preserved.

But is so respectful a visit really possible? Or, rather, could further exploitational and careless visits to the comet be prevented?

We have here a question of invading a wilderness in the sky that we all have been able to see and be inspired by, but not touch. Can humanity lay its dangerous touch on Halley and not have it become deadly to the comet? We think not only about man's exploitation of natural resources and boundless production of pollution. We think also about overzealous experimenters and knowledge-vampires who may end up with less of even purely scientific information than they could have had because they will play a role in the destruction of the thing to be studied or (more likely) no role in its preservation. Like a character in one of George MacDonald's stories, they care nothing about the thing itself, only about knowing it.

Yet perhaps there is hope. It may be in—or perhaps beyond—a comparison of Halley's Comet with Yellowstone National Park. I write "beyond" because we should hope that the preservation of the comet could be made far more successful than that of Yellowstone. The mobility and mighty activity and the smallness of this island in the sky, the comet's nucleus, are all features that decrease its accessibility to invasion. On the other hand, as humankind becomes more adept at space travel, these barriers will of course weaken. We have only a breathing-space in which to take steps to preserve Halley's Comet. It is a breathing-space one or two Halley-periods long, one might think, but we must remember that the interest in Halley— even after the astonishing revelations of the 1985–86 return—will eventually fall (is already falling) into a dreaming sleep. Right now, even though the months of its visibility to the public have long since passed, we must take the time to think about the comet's saving. The time will make itself for people who were entranced by the comet (even while being disappointed!) at this past return. Action to preserve Halley's Comet might be the ultimate answer to the question "What can I do to find a worthy expression of the delight and wonder I gained from this comet and my experience of observing it?" You might even say that this is your way of helping to pay to Halley's Comet a debt of gratitude—even if you did not see it in 1985–86, your parents or grandparents or great-grandparents may have in 1910, and your fellow humanity has hundreds of Halley's visits, back through all history and far before, to be grateful for.

And this is where Yellowstone Park comes in. The lesson to be learned from Yellowstone for lovers of the beauties of the solar system was pointed out in a brilliant letter by the space artist Ron Miller in the March/April 1984 issue of the Planetary Society's publication *The Planetary Report*. Miller proposed that we should already begin efforts to get certain places of outstanding aesthetic and scientific importance in the solar system protected by an internationally respected code of law. When have the opinions of scientists, artists, or naturalists ever prevailed against the slashers and burners and users, or the calculators of strategic arms locations in the

struggle for political and military power? Not often or wholly, but sometimes—and maybe increasingly as our century has become more aware of
the deadly alternatives. As Miller suggests, one of the keys to victories
against exploitation has been foreseeing its possibility and acting long before
exploitation became practicable to raise the world's conscience and opinion
against it.

The story of Yellowstone is a fascinating example. The region was unknown to white men until amazingly late in American history. The stories
of the Indians were deemed only fancies until the first white explorers found
the area and discovered its truly awesome beauties. There were few natural
wonders known that could not be found here in greater size, beauty, or
grandeur than almost anywhere else in the world. Rugged ranges of mountains were its bones, snow-capped and in many places as sharp-peaked and
powerfully thrust as one finds in the imagination of mountain-ness but
seldom in the real world. The high plateau of the region had once been a
volcano which rivaled at least the width of the Martian giants, and now its
thousands of square kilometers were thick with the straight upright or fallen
lodgepole pines and a profusion of wildflowers in meadows by swift rivers
running out of gorges as much as hundreds of meters deep from the snowy
mounts over waterfalls of all shapes and sizes up to that of the greatest
torrent several hundred feet tall. Yet in addition to these known features
of greater proportions more sharply drawn, the explorers walked in utter
astoundment at the marvels of a kind unlike any other in America or even,
in such number and scope, in all the world. Stains of yellow and pink and
other hues were splashed around the spluttering, steaming pools of hot
springs and other thermal features with the harsh, brisk, primeval smell of
sulfur all about, while at every regular or irregular interval imaginable,
columns of steam shot up from geysers (like Halley's jets!) to shine in the
sun. And in the midst of the September-to-June winter when yards of snow
filled the plateau, putting the strange and rich assortment of animals and
plants to hibernation or to the banks of the hot springs or to the test, the
geysers' plumes were cresting into clouds of ice crystals in which there played
in the $-40°$F. (and C.) air the peerless displays of glowing colored arcs,
circles, pillars, and patches of halo phenomena.

The explorers returned to the reality of Washington, D.C. (or perhaps
they had left reality when they left Yellowstone). They returned from that
other world of fierce and alien grandeur, of beauties so wild that death was
not far away, and they reported what they had seen. We may well believe
that it took more than their testimony of wonder, it took hard work to bring
about what then came to pass. The canvases of the painter Thomas Moran
must have been a great help. But whatever the specifics of what they did,

the important thing is that they did it early, before plans for exploitation could be formulated. The Yellowstone region, as large as a small state, was set aside almost immediately to become the first "national park" in the history of the world. The Act of Dedication was signed by President Ulysses S. Grant on March 1, 1872. It spoke of regulations that "shall provide for the preservation from injury or spoliation of all timber, mineral deposits, natural curiosities, or wonders wthin said park and their retention in their natural conditions."

Anyone who has visited Yellowstone in summer knows that in front of the portals to the park sits a dismaying array of tourist appurtenances, that the main roads and some of the lookout points of the park are swarming with people who appear to want a pre-packaged presentation of Yellowstone that lingers longer over postcards and snack bars than waterfalls and forests. But anyone who has visited Yellowstone also knows that just a few minutes' walk off the beaten track, as the rumble of the tour buses recedes and becomes inaudible, there remains a solitude in which to lose oneself utterly in the forests, rivers, mountains, lakes, gorges, hot springs, and cool air—brisker than anywhere else in the summers of the contiguous forty-eight states. Almost all of the wonders have been preserved, and their essential integrity is intact. This unique wilderness of the Earth has survived and is surviving.

If committees of astronomers and other scientists, of artists (spiritual heirs of Thomas Moran), and of naturalists (for nature is in the sky, too) can be organized and can agree on areas in space which we cannot afford to ruin, the precedent will be set, and further discussion can flower. The rings of Saturn, the volcanoes and Valles Marineris of Mars, and the nucleus of Halley's Comet will all come under a prohibition against exploitation that tomorrow's political leaders, even if not much less belligerent and short-sighted than many of today's, will find difficult to overstep.

One expression Thoreau used to describe the top of a mountain in Maine comes to mind when considering the deepest power of Yellowstone. Yellowstone seems like a vast piece of *unhandselled* earth whose sharpness and potency even time has been unable to wear down. Its paradise of harsh, uncompromising, savage beauty so seemingly like the primeval Earth, almost unearthly (even in the unbreathable gases), could be a reflection to the future of what similarly harsh, alien, and beautiful wonders of the solar system will be found to be—and should be kept. Comet nuclei, even somewhat worn ones like Halley's, are thought to be the most pristine and primeval of the solar system's material available to us.

Presumably no wildlife exists on Halley's Comet—no trout flash in its streams (streams turned to vapor before they can exist in the vacuum of space), no snowbirds even sing cheerily on ice-clad branches of trees on its hills. There is no breath there to be safely taken which is part of the comet's own. But, on the other hand, no wilderness of Earth has played such a number of roles in history or been so capable of varied interpretations in its giant mystery as Halley's Comet. As alien and unlivable as the environment itself may be, it has a connection with humanity which is powerful and universal in a way that perhaps nothing else we know is: that once-a-lifetime every lifetime appearance that stirs us to look up, and across the centuries.

The comet means very much to humanity. In the previous chapter, I wrote that it was, like life itself, a trouble and a treasure. What additional meaning can we find in it, especially unique meaning or meaning like other very great things—even greater things, central to humanity?

We know now that maybe tens of thousands or a few hundred thousand years back, maybe even before a human kindled a fire—of humanity as well as wood—this comet, and probably it alone, was the Visitor, and the Witness to every full-span human life (a kind of life so terribly rarer in those days).

Another possible meaning for the comet lies in an unusual analogy. The most devoutly religious poets—Dante, the Pearl poet, Milton, Donne, Herbert, Hopkins—did not hesitate to find analogies for Christian principles and even Christ himself in the most mundane and ostensibly dissimilar things. So I will not hesitate to point out that much of humanity in the past 2,000 years has awaited a mystical Second Coming at the end of (this stage of) humanity's life—just as, in a much smaller way, many people who saw Halley's Comet with the special wonder of childhood, have waited for the comet's return at the end of their own earthly lives. The comet's return seemed to them as certain as Jesus' to Christians. In a much lesser sense than any great holy man (let alone the Son of God, whom Christians deem Christ to be), yet in a very important sense nonetheless, Halley's Comet is the light—one special light—of the world. Halley's Comet is not the mediator for mankind which Christ is; the comet does not come because of or for mankind, though it may come because of the Earth in a very literal way (the highly speculative idea that Earth had a crucial role in Halley's capture into its present orbit). But both Christ and Halley's Comet have a relation to mankind which is intimate in a respect that, despite their unearthly or more-than-earthly nature, makes them truly mankind's—and mankind theirs. If any Christian should find offense in what I am saying, he or she should not, because I am not suggesting that Halley's Comet could in any way be a cause for such spiritual awakening or rejoicing as

Christ is. Rather, I am suggesting that the Christian's mind is glad to find in yet another place (the sky once in a lifetime!) a reminder of the center of his or her religion—an unquiet mountain of dusty ice in space somehow giving humanity each generation a small beautiful glimmer in reflection of Christ's and God's light.

Lest I stir up controversy or drive away readers of all religious persuasions, I will conclude by considering what might seem a more secular line of thought about the meaning of Halley's Comet. Though complex, this final analogy, may bring us closest to understanding that in Halley's Comet which most strongly stirs in us response or resonance of a similar part of our own nature. Halley's Comet is not only mankind's comet but also a comet in the following way profoundly like mankind.

A deep part of us longs for wilderness and "the unharvested sea," the West or Wild of Henry David Thoreau, the West of J. R. R. Tolkien beyond the Sundering Seas and "the music of the last wave falling." Comets—especially the brilliant long-period ones—are the very forms of such desirable wildness let loose in the heavens. This wildness is not out-of-control-ness, for it has an order of its own—its very own, and not ours. Halley's Comet is not one of the long-period comets, appearing for us for the first time, or one of the very short-period ones, worn to dimness and overfamiliarity. Guy Ottewell has said that it is the tamest of the wild comets and the wildest of the tame. But I am saying it is another kind of link: a unique link between wildness (in the more profound sense found in Thoreau) and civilization, possessing attributes of both. In fact, Halley's Comet is not only in this respect reminiscent of humans, but especially of eminently civilized yet wilderness-seeking individuals like Thoreau himself.

Thoreau said that his walks from his cabin at Walden Pond (or Concord) were extravagant saunterings that were like the parabola or hyperbola of a comet's orbit with his house at the Sun's position and the open end of the curve of course facing the west and wilderness. He did not say, yet probably knew, that it is questionable whether any comet's orbit is truly hyperbolic or parabolic. Like comets', his orbit in the vicinity of the sun (his home) was virtually indistinguishable from that of the open-ended curve, yet was not such, for his walks always brought him home eventually to Walden at least, his outpost on the edge of the civilization of Concord. But Thoreau distinguished himself with spirituality and reason from the men of lower wildness like the Canadian woodchopper in *Walden*. Tolkien has argued that the writer of effective fantasy or mythology (whose heart is high, spiritual wildness) requires a very high level of knowledge about

Henry David Thoreau, J.R.R. Tolkien, and Halley's Comet

the common world of ordinary reality we live in, and of reason (the fantasy writer should be the master of reason, neither its slave or its evader).

While this analogy between these two writers and between them and what Halley's Comet means to us is limited by a host of numerous and illuminating dissimilarities, it is itself a road for thought that leads us to some interesting places. Halley's Comet alone of all bright and strongly wild comets has returned to us—not just once but repeatedly—to teach us about such wildness; yet at the same time it has had our culture and history adhere to it. Like a myth or fairy tale in Tolkien's cauldron of Story, Halley's Comet has become smoked with human thought, steeped in human imagination, simmered in the juices of both Story and History. Tolkien says that the most interesting myths or fantasies are those that deal with meetings or crossings on the borders between our world and the Other one, the world sometimes called fairyland and which I am here calling the world of high wildness. A mortal wanders into that Realm of Wonder, Desire, and Peril, and it is his own lack of wisdom about his heart or soul that puts it in the peril of falling to evil or of suffering not constructively but destructively the grief of bereavements (lovers of the heavens are bereaved for life of Halley, and its passing brings to mind bittersweet memories of those who did not live to see it or will never see it again—a far greater bereavement). Thoreau pines very constructively indeed for all of us about his realm of wildness that one can never quite possess save only in the clean, clear desiring itself.

Perhaps Halley's Comet is a wanderer from that realm beyond our own common reality which our increased knowledge and ability to visit will bring to resemble us even more in our attempts to reach and understand it. That is to say: the comet will become like today's (and tomorrow's) most desiring Halley-watchers in being mostly within the common "world" of our experience but still, in its mysteries (like our own), that special kind of venturer into wildness. It will be venturing over the border once in a while back to its old home of Faërie, or sauntering for a while west from its new home at Walden (where we meet it) into the holy and again Halley wilderness.

15

Walking to the Stars

IN THE BEGINNING of this book I wrote that all we need to see a myriad of wonders that others miss is the two eyes with which we were born. Not thousands of dollars' worth of telescope equipment, not computers, not a lifetime of experience, not perfectly clear skies—though all of these things are wonderful when you get them if, and really only if, you have first used your eyes.

That is what I wrote at the beginning of the book: your two eyes are the first and only things you need to get started in astronomy.

But perhaps I was wrong! Perhaps there is something you need even before the eyes.

Perhaps it is *the feet*.

Science-fiction writers have long imagined spaceships that would take us to the stars, but the Einsteinian ultimate speed limit of light scorns their fancies about breaking it, and otherwise the journey takes too long—more than four years at light-speed even to reach the nearest star! Yet I say that the first, and in one key respect best, way we can reach the stars is not with hyperspace-drive, warp engines, or hydrogen-scoop time-dilation space-craft. It is with *the feet*. We can *walk*.

Now I am certainly not belittling the ability of the eyes, nor the goal of a working starship (which I not only hope but believe we will some-century build to carry us bodily and gloriously to the stars). But we must walk to the stars before going there any other way. What I mean by this jesting yet profoundly serious statement is essentially what C. S. Lewis meant by one of his comments in an essay on science fiction. No man could find an abiding strangeness even on the moon, Lewis wrote, if he could not also find an abiding strangeness in his back garden. My point is just that: it will not do the human spirit any good to visit the stars by spacecraft if that spirit has not first had the experience of a walk away from buildings and manmade lights to the stars—and imaginatively right out among them. The future space traveler who has not first made the simple walk of wonder will find his journey beyond our solar system ultimately a disappointment. After an initial thrill, that person's jaded mind will learn that other star systems, the stars themselves, were no closer to what he had lost than his home planet was. Like his homeless home, they would be empty, a nothing-but, ashes and dust.

The most important practical message which has been written (at least between the lines) on every page of this book is as follows: *Get out and observe!* The Starry Room is waiting for you.

To stress this message one final time, and in it tie together the many strands of this book, let us take in the words of this chapter a last walk together—in the hopes that the next one you will take with your real feet, and with your eyes not to the page but to the heavens.

If you live in or near a city, as do most people today, I concede that you may have to use a vehicle before you use your feet. But it need not be a spaceship. The oldest and most ailing jalopy will usually suffice to drive you all the way to the stars!

Moon and planets and glimpses of many other things can be had from the midst of many cities, but to become acquainted with the starry sky on its own terms you should make that drive out of town. To take it even just once a month or a season may be enough if you use it well. When you have left the glows of the cities far behind, then your car door will swing shut behind you, and your feet will take you on those few most important of all steps in astronomy: the steps outside under a gloriously starry sky.

If you live in the country, of course, the universe might be just down the road, right in that field—or even just out the back door of your home. All you have to do is walk.

On this night we are star-adventurers from a city, and we have driven far out into the country. The inside light of our car winks off as we close the door behind us. Twilight has ended, and though a thin sliver of moon is casting some light in the west, what we first notice is how dark it seems! But this is the darkness which mankind has lived with for countless thousands of years and usually found more beautiful than fearful. A biologist would tell you that the human body was meant to experience such darkness; a poet or preacher or naturalist or devoted star-watcher would tell you that the human spirit was also meant to—and meant to find wonder in it.

Within a few minutes you discover that you can see your surroundings far better than you initially could and that ever more stars are becoming visible. (The stars! They are *there,* and everywhere. Have you ever seen them, even in a dream, so numerous and rich and present and alive?) To calm part of yourself down a bit, you think about observational facts and techniques. You are allowing yourself some time for dark adaptation of the rods in the retinas of your eyes. In about twenty minutes your ability to see faint objects will be even better—though of course it will never be near its very best until that moon sets. But already you keep looking in amazement at how richly sprinkled the sky is with stars—not the feeble, isolated incidents of the city sky but individuals in throngs and crowds patterning all corners of the heavens. Even the very faint ones seem more prominent against the dark sky than the brightest did from the midst of a large city. And the brightest stars are now living hearts of companionable fire. Despite the moon, you are even using the technique of averted vision to test what are the faintest stars you can see in various constellation patterns. And as you try to trace those patterns, the millennia of myths and legends about them come weaving back into your mind.

But you soon have second thoughts about neglecting the fair natural satellite of Earth. You notice some "Earthshine" illuminating the dark part of the lunar disk. You try to see fine detail on its familiar face. And in the state of mind which this effort of walking and this environment of stars has put you in, you find that face one you have always known yet suddenly also new, fresh. Perhaps you have seen a lunar eclipse before. This night's view may remind you of an extra thrill you got then—realizing that even *without* the eeriness and eventfulness of an eclipse, the moon's face is utterly beautiful and fascinating. Certainly on this night, no eclipse is needed.

The amazing moon is almost always there; you just have not looked. Even the sky, largest of all sights, has escaped your attention on most nights and days—and city lights do not interfere much with fair views by night of the moon or at all with fair views by day of countless different kinds of halo phenomena and rainbows. Admittedly, to find many of the halo effects and rainbows often you need a little special knowledge (as in Chapters 4 and 5 of this book). But many people do not seem to be able to find the sky itself these days. It becomes far more difficult when there are so many walls and ceilings and buildings about you. But if such walls are not to become those of a kind of prison which your mind itself imitates, you must seek the sky— and maybe its most far-flung or highest of outposts: the stars.

Suddenly your thoughts are pierced by the streak of a meteor. The time is still rather early in the evening to see many "shooting stars," but before you have ceased wondering about the first, another breaks the black of night with instant, glowing, passionate eloquence-in-wildness. No major meteor shower is expected tonight, but the sky is very clear (you have correctly chosen excellent weather for the night of your drive to the stars), and away from city lights numerous meteors are always to be expected. But then the unexpected, the most brilliant meteor of the evening passes majestically, flaring to a brightness greater than Venus's before disappearing in a still-brighter burst. If it was fairly low in the sky it must have been even brighter for observers many dozens of miles off in that direction. If that was the case and you are able to recall its path in the sky rather well (despite your shock), then you should consider reporting it so that searchers in a neighboring state (perhaps) might find a piece of space rock or iron, a meteorite, from it. Your chances of seeing such a fireball meteor are at their greatest at this time of the night, but you were still very lucky. The fireball may turn out to be the most unusual and certainly startling event of the night, and the one you most precisely remember. But there are hours' and whole seasons' worth of constellations and countless rich light-years yet to go!

The moon is setting behind a treeline, and you now admire all the more a spread-out line of two planets and one bright zodiac star extending to the upper left from it across the southwest sky. You must remember to watch the moon's progress the next few nights especially to see how close it gets to each of these points of light, how the configuration changes.

Are you sure which planets you are seeing, and whether that one bright point of light is a star? The twinkling of the one is an obvious contrast to the steady shining of the other two. It must be a star; they, planets. One of the planets is far brighter than any star—you remember that Venus is now visible only just before dawn, so the planet you are viewing now has to be the other very brilliant one, Jupiter. The other planet would have to be either Mars or Saturn (Mercury would never be visible so high or so long after twilight). But Saturn, you know, will not be near Jupiter in the sky for many years. So the dimmer planet must be Mars. And indeed you do notice that it looks a little ruddy.

You wish you could enjoy these sights of the planets in a rich, over-spreading field of stars every night. You will have to try to get out more often. But even people blessed with country skies and clear climates sometimes have to deal with clouds or too bright a moon. A heavens painted with phosphorescent paint on black construction paper (or its like) on your ceiling seems a more attractive idea than ever, easily worth the few dollars, even worth a lot of work to make accurate. Here beneath the country splendor of the real stars, there is no concern about such painted stars being phony (it being obvious that representations—including the planetarium's—fall immensely short of this reality). The increased motivation to make an indoor sky is the desire to do in even that small way what some awesome power has done with a multitude of suns over the virtually limitless perspective of space visible from Earth (imitation is, in this case at least, the sincerest form of admiration).

But painting an indoor sky of glow-in-the-dark stars might also help you learn your constellations better. While you pondered, the moon has set, and the sky has rather suddenly grown noticeably darker. You are chilled by the beauty (maybe also by the air—button up that coat!). Still more crowds of stars have appeared. You do not know them well enough perhaps to locate the famous "variable stars" and check how bright these are tonight. You do not even know your constellations well enough maybe to be sure if one of the stars you see does not belong there. Could one of those stars be the first naked-eye sight of a fresh supernova which some amateur sky-watcher is greatly aiding science by early discovery of tonight? Perhaps it is even the long-awaited one that will outshine Venus or rival the moon for a few months.

From these exotic ponderings you turn to some solid work at trying to identify a few constellations with a book or astronomy magazine's star charts you have brought (you use a piece of red cellophane over your flashlight so as not to spoil your dark adaptation). If you learn just a few constellations each month, by the end of the year you will have found all the more conspicuous ones, and locating the less-bright ones near them will be little problem once you have that initial framework of knowns to work from. Right now, the richness of stars and possible patterns in them is a cause for bewilderment—wonderful bewilderment.

Casually you turn to another direction for the first time since moonset—and stop short. There is a glow down at the horizon, and you are almost certain it is that of the city you came from. How quickly will it advance up the sky in the next year? You cannot let this threat to the heavens ruin all your enjoyment of them this night—but neither can you forget "light pollution" or settle for doing nothing when chapters like "The End of the Stars" in this book tell how you can fight it and begin to turn the tide against its wave of star-killing and money-wasting.

If you did not bring a telescope or binoculars (those amazingly portable, relatively inexpensive dual telescopes!) and do not know the star clusters, nebulae, and brightest galaxies fairly well, then you are exceedingly unlikely to happen upon a comet tonight. Naked-eye comets are visible at least dimly, at least once every few years from country sites like this, though. Right now, while you are scanning the heavens, many amateur observers with more knowledge and experience than you but perhaps no more equipment than a small telescope or large binoculars are hunting the western sky in these hours just after evening twilight, an especially good time and place for discovering new comets. Every year a small number of amateurs succeed in catching a new comet and having the thrill of getting their name attached to so vast and ancient an object of mystery. If you did see that king—though far from brightest—of comets, Halley's, in 1985–86, then you may have appreciated its elusive but marvelously varied and significant beauties. Whether or not you did, one of the reasons to consider buying a telescope—*after* you have learned the heavens with naked eyes or with binoculars—is to track down Halley's usually dim but always interesting kin.

The thought of Halley's Comet and the long, long wait for its next return reminds you of the immensities of time and space. Those of the several thousand stars now burning all around you are far greater than even Halley's whose lifespan dwarfs men and women's, whose returns measure the passage of whole human generations. But if you have read this book, then you are hardly feeling insignificant beneath the starry firmament. You know that human beings have watched the many corners of this great Room with rapt attention and admiration, being moved beyond themselves in awe and appreciation—then back again to their true selves with a sense of belonging. The creature who has learned so much in this Room is not insignificant, and his delight in what is seen and understood of it makes him not a shipwrecked alien in it or stranger to it but its friend and lover.

The hour is getting late. You look to the east and see that the Earth is rolling you toward a bold new headland of stars. You are on your way to them. Without leaving your feet, without leaving the dear Earth finally (or maybe at all), you are departing to the heavens. You are walking to the stars. Somehow you seem to know not only that you are traveling, but also that you are traveling to what is wonderful—an understanding it would be well to remember in our lives, so that it would there also, as on this night, become true.

SUGGESTED READING/RESOURCE LIST

ECLIPSES

Littmann, Mark, Ken Willcox and Fred Espenak. *Totality: Eclipses of the Sun.* Oxford University Press, Inc., 198 Madison Ave,. New York, NY 10016
http://sunearth.gsfc.nasa.gov/eclipse/eclipse.html
http://www.mreclipse.com/MrEclipse.html

Harrington, Philip S. *Eclipse!: The What, Where, When, Why, and How Guide to Watching Solar and Lunar Eclipses.* John Wiley & Sons, Inc., 605 Third Avenue, New York, NY 10158; www.wiley.com

RAINBOWS AND HALOS

Greenler, Robert. *Rainbows, Halos, and Glories.* Reprint available from Blue Sky Associates, Inc., 920 Seventh Avenue, Grafton, WI 53024; (262) 377-1398 www.blueskyassociates.com

PLANETARY CONJUNCTIONS

Meeus, Jean. *Astronomical Tables of the Sun, Moon, and Planets.* Willmann-Bell, Inc., P.O. Box 35025, Richmond, VA 23235; (804) 320-7016 www.willbell.com

METEORS AND METEORITES

Meteor Trails. Quarterly Journal of the American Meteor Society. AMS, 3859 Woodland Heights, Callahan, FL 32011; www.amsmeteors.org

LIGHT POLLUTION

International Dark–Sky Association, 3225 N. First Avenue., Tucson, AZ 85719; www.darksky.org

HALLEY'S COMET (AND COMETS IN GENERAL)

Ottewell, Guy and Fred Schaaf. *Mankind's Comet.* Available from Astronomical Workshop, P.O. Box 426, Middleburg, VA 20118; 888-432-2264; http://www.universalworkshop.com
Also available from dealers in out-of-print books.

Schaaf, Fred. *Comet of the Century—From Halley to Hale-Bopp.* Springer-Verlag New York, Inc., 175 Fifth Avenue, New York, NY 10010 www.springer-ny.com

GENERAL: ANNUAL ASTRONOMICAL ALMANAC

Ottewell, Guy. *Astronomical Calendar.* Same address as for *Mankind's Comet* above. Also available from the online bookstore of the American Association of Amateur Astronomers: www.astromax.com.

GENERAL: MONTHLY ASTRONOMY MAGAZINES

Astronomy. Kalmbach Publishing Co., 21027 Crossroads Circle, P.O. Box 1612, Waukesha, WI 53187-1612; (800) 446-5489; www.astronomy.com

Sky and Telescope. Sky Publishing Corp., 49 Bay State Road, Cambridge, MA 02138-1200; (800) 253-0245; www.skyandtelescope.com

GLOSSARY OF TECHNICAL TERMS

AND CONCEPTS USED FREQUENTLY

IN THIS BOOK

POSITIONAL MEASUREMENT IN THE SKY. In the heavens, the equivalent of longitude and latitude on the Earth are, respectively, *right ascension* ("*R.A.*") and *declination*. R.A. is measured in hours, minutes, and seconds, there being 60 seconds (s) in a minute (m), 60 minutes in an hour (h), and 24 hours of R.A. to encircle west to east the entire heavens above and below the horizon. Earth's longitude is measured 180° west and 180° east from the meridian which passes through the Greenwich Observatory in England, but R.A. increases (1h, 2h, 3h, . . .) continuously to the east from a meridian that runs through the vernal equinox (the point the sun reaches in the sky at the start of our Northern Hemisphere's spring). Declination is measured 60 seconds (″) to a minute (′) and 60 minutes to a degree (°), with positive (+) degrees running for 90° north from the *celestial equator* and negative (−) degrees running 90° south from the celestial equator (the celestial equator is the strip of heavens which passes overhead as seen from Earth's equator). The *north* and *south celestial poles* (at +90° and −90°, respectively) would be seen overhead from Earth's north and south poles.

There are also ecliptic and galactic systems for positional measurement in the sky. The *ecliptic* is the sun's path in the sky, in reality a projection of the plane of Earth's orbit. It is the midline of the band of sky in which we also find the moon and the planets, the band that passes through (actually more than) 12 constellations and is called the *zodiac*. In the ecliptic system of measurement, the ecliptic itself is the equator, with *ecliptic latitude* running to 90° north and 90° south, and *ecliptic longitude* running eastward from the vernal equinox in degrees up to 360°.

DISTANCES IN THE SKY. Distance in the sky can be measured in degrees (°), minutes (′), and seconds (″) of arc. The moon and sun are roughly 30′ or ½° wide. The bowl of the Big Dipper is about 10° across the top,

258

and 5° between the two "Pointer" stars on its outer side. Most people's fist is about 10° across, their little finger about 1½° across, when their hand is held out at full arm's length.

DISTANCE IN SPACE. Distances in the solar system are often expressed in *A.U.s (Astronomical Units)*, an A.U. being the average sun-to-Earth distance of about 92.9 million miles. Distances between the stars and galaxies are often measured in *light-years,* one of these units being the distance that light, moving at 186,000 miles per second in a vacuum, can travel in one year.

PLANETARY POSITIONINGS AND EVENTS IN SPACE AND THE SKY. When a planet (or other solar system body) is nearest to the sun in space, it is at *perihelion;* when farthest from the sun in space it is at *aphelion.* When a planet or other object is passing due north or due south of the sun or any other object in the heavens, the two are said to be in *conjunction.* When a planet is on the opposite side of the heavens from the sun (as nearly 180° away as possible), it is at *opposition*—it therefore rises at sunset, is visible all night long, and, around this time, is closest to Earth (therefore brightest and biggest in our skies). Only the *superior planets,* those farther out from the sun than the Earth is, can come to opposition. The *inferior planets,* those nearer to the sun than Earth is, reach *greatest elongation* (largest angular separation from the sun in the sky) and, at another point in their orbit, *greatest brilliancy* (a technical term that is almost but not quite the same as, and not quite coincident with, brightest).

MEASUREMENT OF BRIGHTNESS IN ASTRONOMY. Astronomers usually measure the brightness of celestial objects in terms of *magnitude.* The lower the magnitude, the brighter the object (thus a 2nd-magnitude star is brighter than a 3rd-magnitude star, a magnitude 1.5 object brighter than a magnitude 1.7 object). A difference of 1 magnitude equals 2.512 times (this is so that a difference of 5 magnitudes can equal 2.512^5, which is 100 times (thus a star of magnitude 1.0 is 100 times brighter than one of 6.0). A few very bright stars and planets are less than 0 magnitude (Sirius is −1.4, Venus at brightest about −4.6). The full moon is almost −13, the sun almost −27. On the faint side, the dimmest star visible to the unaided eye in good conditions is often said to be about magnitude 6.5, the dimmest in a 6-inch telescope is around 14 (or a little dimmer), and the dimmest ever recorded with sensitive electronic light detectors on the world's largest telescopes is about magnitude 26.

INDEX

Ablations, meteorites, 113
Ahnighito meteorites, 120–121
Albers, Steve, 180, 186
Alexander's dark band, rainbow, 53
Alpha and Beta Centauri, 200
Angular altitude, 89
Anomalous sounds, fireballs, 94–95, 100, 133
Anthelion, 45
Anti-solar point, 51
Archer, Don, 219
Asteroids, 118
    asteroid belt, 103
    Big Four, 148
    as origin of fireballs, 91, 102–103
    as origin of meteorites, 113, 129–130
Astroblemes, 118
Aurora borealis, 13
    color perception, 186
Austin, comet, 84
Autumn, rainbows, 63
Azimuth, 89–90

Bacubirito meteorites, 120, 122
Barringer Crater, 117
Beamon, Carol, 219
Bearman, Harry N., 76
Bennett, Comet, 228
Betelgeuse, 183
Big Dipper, 177, 180
Billet, Felix, 54
Biot, Jean Baptiste, 108, 123
Black rainbow, 66
Blue-sky rainbow, 64
Bolides, 79
Boyce, Peter, 219
Brown, Robert, 124–126
Bruderheim meteorites, 133
Burnham, Robert, Jr., 174

Capricornids, 78
Carbonaceous chondrites, 130
Celestial navigation, animals, 175
Ceplecha, Zdenek, 131, 134
Ceres, 148
Chinguetti meteorites, 120, 121
Circumzenithal arc, 13–14, 59
Clark, David H., 193, 202
Closest rainbow, 54–55
Cloudbow, 58
Collimation, 180
Collins, Michael, 152
Color perception, 183–188
    age factors, 185
    aurora, 186
    Purkinje effect, 185
    rating star colors, scale for, 184
    spectrum and human sight, 184–185
Comets
    and mass extinctions, 118
    and meteor showers, 76
    sight by naked eye and, 189
Countersun, 46
Crab Nebula, 193, 196, 200
Craters, meteorites, 117–118
Crawford, David, Dr., 209, 212, 219, 221–222

Danjon scale, 28–29, 32
Dark Skies for Comet Halley (DSCH), 218–221
Delta Aquarids, 78
Dewbows, 59–60
Diamond-ring rainbow, 67
Dog meteorites, 120
Double rainbow, 53
Double stars, 179–180
    colors of, 184
Double sun, 45

Eclipses, 15
  colors, 26, 28
    brightness scale, 28–29, 32
  dark eclipses, 1601–1982, 34
  of July, 1982, 21–41
    Danjon rating, 32, 33
    compared to December, 1963
      eclipse, 22, 34–35
    and El Chichón volcano, 26, 28, 29,
      32–33
    length of eclipse, 24–25
  magnitude of, 25
  total between 1982–1999, 23, 24
Edberg, Steve, 228
Edwards, Jonathan, 65
Edwin Hubble Space Telescope, 238, 240
Eiseley, Loren, 238
El Chicón volcano, and 1982 eclipse, 26,
  28, 29, 32–33
Ellis, George, 219
Epsilon Geminorum, 180
European Network, 130
Explosion craters, 116
Extinctions, mass, and comets, 118
Eye
  animal sight, 175, 177, 190–191
  averted vision, 170
  brightness acuity, 188
  collimation, 180
  color perception, 183–188
  dark adaptation, 169, 173
  diffraction spikes, 179
  dim extended glows, perception of,
    189–190
  fields of view, 188–189
  limiting magnitude, 170–171, 173–174
  observation, components of, 171–172
  response to light intensity, 167–169
  revolving power, 177
  sharpness of, 177–179
  Troxler phenomenon, 173
  unaided observation, example of, 173–
    174
  visual acuity tests
    double stars, 179–181
    Jupiter's moons, 182
    lunar features, 181–182
    Venus' crescent, 182

Farthest rainbow, 54
Fifth-order rainbow, 53–54

Fireball meteor rainbow, 67
Fireballs, 79 *See also* Meteors
Fleet, Richard, 233
Fogbow, 58
Fourth-order rainbow, 53
Full-circle rainbow, 55

Geminids, 75, 76, 105
Giacobini-Zinner, Comet, 228
Grand curve, 141, 143
Great Meteor Procession, 98
Green flash, 15
Greenler, Robert, 56, 63, 66
Gribbin, John, 138, 140
Griffin, Arthur A., 112

Halley, Edmund, 109–110
Halley's Comet, 7, 174
  Dark Skies for Comet Halley (DSCH),
    218–221
  discoveries/findings about, 229, 231–
    232
  future view, 237–242, 244
  meaning of, 245–248
  spacecraft passing near, 229, 239
  tail sightings, 232
Halliday, Ian, 112
Halo phenomenon
  anthelion, 45
  circumzenithal arc, 13–14
  countersun, 46
  double sun, 45
  mock suns, 43–44
  paranthelia, 45–46, 47
  ring around moon, 12, 44
  seven suns of Hevelius, 46–47, 48
  subsuns, 44–45
  types of, 43
Hawaiian rainbow, 64–65
Haze, rainbows, 54
Hermes asteroid, 118
Hevel (Hoevelcke), Johannes, 46–47
Hevel's halo, 47
Highest rainbow, 56
Hoba West meteorites, 120, 122
Horse and Rider, 180
Houston, Walter Scott, 123, 124–125,
  171, 185, 190
Hurley, James R., 219
Hyades, 17

Infrared rainbow, 63
Innisfree meteorite, 132, 133
International Dark-Sky Association
    (IDA), 222

Juno, 148
Jupiter, moons of, 182
Jupiter effect, 138–141

Kepler, Johannes, 145
Kepler's Star, 193
Kirin meteorite, 115–116
Koch, Edward, 219–220
Kohoutek Comet, 233
Krakatoa, 35

La Paz, Lincoln, 123, 128
Lightning rainbow, 65–66
Light pollution, 206–224
    Dark Skies for Comet Halley (DSCH),
        218–221
    increase of, 210
    International Dark-Sky Association
        (IDA), 222
    outdoor lighting
        low-pressure sodium (LPS) lighting,
            advantages of, 213–215
        principles of, 211–215
    severity of, 207–209
    Walker's Law, 208, 209
Limiting magnitude, 170–171, 173–174
Lost City meteorite, 131
Lovejoy, Terry, 232
Low-pressure sodium (LPS) lighting,
    advantages of, 213–215
Lunations, 34
Lupus supernova, 196–198, 200
Lysol bow, 64

Man meteorite, 120
Marsden, Brian, 77
Martinez, Tom, 219
Mascart, Eleuthère, 53
Mbosi meteorites, 120
Meeus, Jean, 25
Meinel, Aden and Marjorie, 66, 186
Meisel, David, Dr., 98, 102, 103, 133
Mercury, elusiveness of, 12–13

Meteorite Observation and Recovery
    Project (MORP), 130, 133
Meteorites
    ablations, 113
    building struck by, 127–128
    chances for falls, 111–113
    classes of, 114–118
        asteroids, 118
        bright naked-eye meteors, 115
        fireballs, 115
        micrometeorites, 114
        telescopic meteors, 115
    composition types, 113
    famous meteorites, 120–123, 131–132
    historical view, 108–109
    humans struck by, 128–129
    iron, use of, 110
    measurements related to, 129
    meteorite hunters, 123–127
    meteorite shower, 116
    Meteor News, 125
    origins of, 113, 129–130
    properties, and fall of, 110–111
    tracking of, 130–134
    types of, 130
    vaporization types, 113
    worship of, 109
Meteoroids, 73
    streams, 73
Meteors
    bolides, 79
    fireballs, 79
        asteroidal origin of, 91, 102–103
        of August, 1982, 83–96, 98–105
        fragmentation of, 101
        of long duration, 97–98
        measurement difficulties, 91–92
        1982–1983 sightings, 95–96
        SEAN *Bulletin*, 87, 95, 96
        sounds from, 86–87, 94–95, 100
        sparks from, 86
    Great Meteor Procession, 98
    measurement methods, 89–90, 92
    meteor showers, 12, 73–74
        and comets, 76
        Delta Aquarids, 78
        major annual showers, 73
        Perseids, 73, 75–78, 79–81
    meteor storms, 73
Metonic cycle, 34
Micrometeorites, 114
Miller, Ron, 242

Millman, Peter, 134
Minnaert, M., 45, 46, 55, 57, 179
Mistbow, 58, 65
Mizar and Alcor, 180
Mock suns
    causes of, 44
    characteristics of, 44
    rainbows of, 66
Monnig, Oscar, 124–127
Moon
    bright, and observing night sky, 10
    crescent, 15
    features for testing visual acuity, 181–
        182
    first quarter, 11
    full moon, observable lunar features,
        10
    terminator, 10
Moonbeams, 30
Moon rainbow, 64, 66
Moore, Patrick, 180
Morito meteorites, 122–123
Morning rainbow, 56
Morris, Charles, 228, 232
Mountaintop, observation from, 174
Mt. Katmai, 35

Nelson, Richard, 65
Newburn, Ray, Jr., 219
Nininger, Harvey, 108, 123–124
Noctilucent clouds, 114
NORAD, 102
Northern Lights, best time for, 13
Nuclear winter, 118
Nu Draconis, 180
Nullabor meteorites, 120

Observation of night sky
    after cold front, 11
    blackened tube, use of, 174
    and bright moon, 10
    cloudy nights, 11–12
    good/bad "seeing," 11
    planets, 12–13
    See also Eye
O'Keefe, John, 98
O'Meara, Steve, 174
Opposition, planets, 12
Orion, 17
Orionid meteors, 7
Ottewell, Guy, 79, 80, 141, 218

Paired opposites, planets, 143
Pallas, 148
Paranthelia, 45–46, 47
Parhelia, 44
Parhelic circle, 44
Parselene, 30
Peary, Robert, 108, 120–121
Pena Blanca meteorite, 126
Penumbra, 22, 27, 30, 37
Perseids, 73, 75–78
Pickering, W. H., 181
Pioneer spacecrafts, 142
Plagemann, Stephen, 138, 140
Planetary nebulae, 185
Planets
    concentrations of, 138–141, 148–153
    grand curve, 141, 143
    Mercury, 12–13
    oppositions of, 12
    paired opposites, 143
    planetary conjunctions, 137
    rare planetary events, 148–153
    in retrograde, 144
    space program exploration, 141–143
    Triple Conjunction, 145–148
Pleiades, 17
Pliny, 109
Prairie Network, 130, 131
Pribram meteorite, 131
Primary rainbow, 53
Purkinje effect, 185
Purkinje, Johannes, 185

Quadrantids, 75, 76

Rainbow pillar, 63
Rainbows, 13
    circumzenithal arc, 13–14
    types/descriptions of, 53–67
Red rainbow, 55
Redwing Creek meteor, 118
Reflections, and rainbows, 60–63
ReVelle, Douglas, 131
Ridwan, Ali Ibn, 198
Ripert, M., 122
Robinson, William, 221
Rosenblatt, Roger, 223
Rose of rainbows, 54

Schiaparelli, Giovanni, 76, 77
Schwarz, Gunther, 131

SEAN *Bulletin*, 87, 95, 96
Searchlight rainbow, 58
Secondary rainbow, 53
Sekanina, Zdenek, 117
Setting rainbow, 56
Seven suns of Hevelius, 46–47, 48
Sherlin, Jerry, 219
Sirius, 199
Sky, indoor, construction of, 158–164
Snow, rainbows, 56
Star of Bethlehem, 145
Stars
    shooting stars, 73
    twinkling phenomenon, 198–199
Stephenson, F. Richard, 193, 202
Storms, rainbows, 56, 58
St. Lawrence's Tears, 80
Subsuns, 44–45
    characteristics of, 44–45
    subsun dogs, 45
Sun dogs, 44
Sun pillar, 44, 45
Sunset
    green flash, 15
    red rainbow, 55–56
Sunspots, 13
Supernova
    Crab Nebula, 193, 196, 200
    historical view, 193–195
    Lupus supernova, 196–198, 200
    occurrence of, 202
    twinkling and, 199
Supernumerary arcs, rainbows, 57
Swift-Tuttle Comet, 76–77
Swimmer's rainbow, 53

Talithids, 105
Tektites, 130

Telescopes
    Dawes' limit, 180
    rich-field type, 189
    speed of, 189
Tertiary rainbow, 53
Tornadoes, rainbows, 57
Triple Conjunction, planets, 145–148
Troxler phenomenon, 173
Tunguska event, 117
Tycho's Star, 193

UFOs, 45
Umbra, 22, 27, 37

Valeč meteorite, 133
Van Dyke, Richard, 208
Variable stars, 188, 194
Vega, 195
Venus, 13
    crescent, sighting of, 182–183
Vesta, 148
Victor, Robert C., 145
Volcanic eruptions
    and eclipses, 26, 28, 29, 32–33, 35
    and rainbows, 65
Voyager spacecrafts, 141–142, 143, 146
Vredefort Ring, 118

Walker, Jearl, 54, 58
Waterfall rainbow, 58
Whipple, Fred, 110, 229
White rainbow, 58
Widmanstatten structure, 130
Willamette meteorite, 121–122
Woman meteorite, 120